John William Bennett

A Breed of Barren Metal

Currency and Interest, a Study of Social and Industrial Problems - Vol. 1

John William Bennett

A Breed of Barren Metal
Currency and Interest, a Study of Social and Industrial Problems - Vol. 1

ISBN/EAN: 9783337144548

Printed in Europe, USA, Canada, Australia, Japan

Cover: Foto ©berggeist007 / pixelio.de

More available books at **www.hansebooks.com**

A BREED OF BARREN METAL

OR

CURRENCY AND INTEREST

A STUDY OF SOCIAL AND INDUSTRIAL PROBLEMS

BY

J. W. BENNETT

"First freedom and then glory; when that fails,
Wealth, vice, corruption, barbarism at last."

CHICAGC
CHARLES H KERR & COMPANY
175 MONROE STREET

CONTENTS.

5

6 CONTENTS

PREFACE.

I GIVE this book to the public in the hope that it may aid in bringing about a re-examination, in the light of modern intelligence, of one of the principles lying at the foundation of our industrial and social organization. If the effort shall have any influence in bringing about such a consummation, its mission will be accomplished. The foundation must be secure if the fabric is to stand, and I believe that finance and industry now rest on an insecure foundation. The line of argument pursued in this work has never, to my knowledge, been presented by any other person. The reason of the wrong has never before been fully explained. If I am mistaken, he who points out the error will confer a favor on me and thousands of other seekers after truth. But we want reason and not dogma.

The practical application of the truth here presented would be so far-reaching in its effect as to be revolutionary. But it would be a peaceful and beneficent revolution, doing wrong to no one, and bringing justice to oppressed millions.

In the work here presented I have endeavored to call things by their right names. No social wrongs are impersonal, and in denouncing or merely pointing out a wrong those who gain by such wrong must expect to receive some of the censure. I criticise classes for what they do, not for what they are. If the average day laborer who tries to keep body and soul together on a dollar per day should change places with the man who

11

makes a million a year, that laborer would be just as grasping and unscrupulous in money-getting as he. The wealthy are not villains, nor the poor oppressed saints. But that is no reason why the wrong practiced by one class upon the other should not be condemned. We are all as good as we know how to be. I do not believe that any being is perversely or malignantly bad. The trouble is, we are too ignorant to recognize the right and do it. I think that the words of the "Melancholy Dane" should be changed into "blundering fools all." Ignorance is the root of all evil; perfect knowledge is perfect morality.

Then it is necessary to look all questions square in the face and try to learn the truth about them. I hope that I have done this; it is what I have attempted.

The key to the interest question, the topic which I discuss, is the currency of the nation. For that reason I have added a discussion of the currency question and set out a plan for a scientific currency. It is practical and suited to the times. To all liberal-minded persons, it is worthy of most careful consideration. The idea of using wealth in the process of exchange as the basis of a currency system is old. Holland did it successfully. The means of applying that principle are my own. I offer the work to an earnest, truth-seeking public, in the hope that it will further the cause of justice and truth. Examine it. It will repay the trouble.

J. W. B.

A BREED OF BARREN METAL.

CHAPTER I.

INTRODUCTION.—The importance of studies in sociology—Man a social animal—The correct method—Social and economic, compared to physical science—Natural law the foundation of both—All science speculative—Happiness the object of existence—Happiness development—Must be general—Ideal state is happiness to all—In sociology as well as art and science the ideal is the standard—The ideal is that which corresponds most closely with natural. law—Natural law must be obeyed—The first business of social and economic science is to find underlying natural principles—Ethics the foundation of these sciences—Present practices and institutions to be tested by ethical axioms—Unsatisfactory results to be explained by economics—Knowledge for the use of man.

The most important chapter in the book of nature is man.

THERE must be an excuse for all things. My friend the artist tells me why he painted this picture and not that. My scientific acquaintance has the best of reasons for following his particular line of investigation. My clerical friend can give a reason for every sermon, and I feel that I must give a reason for this book. I find it in the condition of fellow men about me. The lot of the average human seems a strangely unhappy one. Why? is an interesting and important study. This volume is intended to wrestle with that problem, from the material standpoint. It will try to discover the why of man's material discomfort. If that is discovered, the discovery will doubtless lead to a ·clearer understanding of the psychological questions which confront us. From the tangible we will try to arrive at the intangible and to throw light on some of the all important questions which harass social man. It is because these problems are unsolved, that this book is undertaken, for I still believe,

"The proper study of mankind is man."

The epigram is as true to-day as when written by

13

Pope or practiced by Byron or Shakespeare. The field is unlimited, the possibilities unbounded. The relations of man to man are less understood than any other branch of knowledge. A correct knowledge of what these should be is the ultimate end of all mental effort. The humblest worker in the field is engaged in the grandest of occupations.

Man is essentially a social animal. All that he has, all that he is, comes from association with his fellow men. The most profitable study of man, then, is in his social relations. As we learn the physical laws of the universe by studying the relations of things, so we may learn the social laws of mankind by studying the relations of man to man. If we would profit by the study of the physical universe, we must learn its correct laws, the great principles on which are builded the harmonies of nature. It is the study and determination of these that has enabled man to harness steam, to chain lightning to his triumphal car, and count and measure and weigh untold worlds. It is by putting himself in harmony with nature's laws that man has made all material advances. Just in the same way we must learn and put in practice, the great natural laws of harmony between man and man. If we would make social and moral advances, we must put ourselves in harmony with the eternal laws of justice on which all stable social and political institutions must be founded.

Sociology and political science have more to do than to catalogue the facts of the rise and fall of prices under this or that condition, or to measure the influence of competition on this or that branch of business. There are great laws to be discovered in the social as in the physical world. We have still to formulate the gravitation and conservation laws of sociology, the foundation principles of social relations. Fortunately for us, Galileo, Kepler, Newton, Laplace and the thousands who have lived before and after them, thought that discovery was a part of the province of science. They did not content themselves with motiveless classifying of cold unrelated facts. Neither should the devotees of social science. They should try to lay a foundation of principles, and they must find these principles in nature. From that which is they should determine what should be.

Happily or unhappily, man can for a time disobey moral law, provided he be content to pay the penalty, and what he does is no criterion of what he should do. He may fail through ignorance to take advantage of nature's social laws, as he has failed and still fails through ignorance to apply the physical laws of the universe. To dissipate this ignorance is the business of the sociologist of to-day.

The foundations of sociology go to the very bed-rock of society; its problems are the problems of social and individual existence. If we can determine why we live and labor, and why we organize society, we have the principles for determining how we should live and labor and how we should organize society.

Looking at the problems of human life by the dim light of man's flickering reason, we are led to conclude that happiness is man's aim and object here below. The pursuit may be conscious or unconscious, the goal may be sought in this world or the next, but ask whom you will, you will find that his energies are bent in the pursuit of happiness. All happiness is relative, not absolute. Experience teaches us that the greatest positive happiness lies in the fullest development of all of man's natural powers and capabilities and the fullest enjoyment of these developed powers. It is this and not freedom from trivial pain which men seek. Every individual among the myriads born to mother earth has a right to this free and untrammeled development and enjoyment of the powers which he possesses equal to that of any other of earth's children. There can be no real, just or lasting happiness unless every human being is included. Our reason revolts at the idea of some being born the puppets of others. In practice, both classes are unhappy.

The ideal state of society is that state which secures all the greatest measure of happiness here below, and any state of society or any institutions, political or economic, which tend to develop a few at the expense of the many, are founded on wrong principles—principles which contradict the natural underlying law of all human society. For society is organized to aid men in reaching the ends to which they all aspire. It is a means of securing happiness to man.

While an ideal state of society is not easily attained, it is to the ideal in sociology as well as in every other branch of knowledge that we must look if we would perform anything worthy. It is the divine ideal in the sculptor's mind, carefully and painfully wrought out under his chisel, which gives the enduring beauty to the piece of rough, coarse stone. It is the sublime ideal in the poet's soul which, fashioned into words, thrills and ennobles and elevates the human heart. All art strives toward the ideal, all science toward the perfect, and it is only as they approach this goal that they produce anything of practical value. The deduction of law is the end and aim of all classification and observation. If we would determine how far they are adapted to the ends for which they are intended, institutions must, then, be tried by ideal standards. We must not expect to find them perfect, but we must find them tending to perfection and not in the opposite direction; otherwise they are founded on false principles and certain to fall.

The natural laws under which men live are superior to man. As part of the universe, man is governed by the laws of the universe. His laws must be in harmony with the laws of the universe. If they are they are aided and strengthened by natural law, if they are not they are annihilated and their supporters crushed. There seems little doubt that the laws of the universe are working for good, but whether for good or bad, they must be obeyed. In their overmastering might they are forcing us onward to some distant goal, to some great beyond, and if we do not wish to court annihilation we would do well to place ourselves in harmony with nature's onward march. We may then use our energies in enjoying our surroundings and developing our powers, instead of wasting them in resisting the inevitable. The smallest natural law can no more be resisted with impunity than the mighty current of Niagara.

If, then, we are ruled by an overpowering force; if grooves are set and an invisible but inexorable power is driving us along these grooves, our first business is to locate these grooves and see that we keep on the track. In other words, we must discover the natural laws which rule us as a social body, and see that our social fabric is reared in harmony with these natural laws. We should find out what is, only for the purpose of knowing what should be. When a principle is once clearly established as a natural law, all other laws must bow to that principle.

Ethics, the science of correct human relations, is the natural foundation stone of the whole social fabric, and I propose to test with the principles of ethics some of the principles applied in society.

As the mathematician finally puts the test of axiomatic truth to every principle of his science, and by that test determines whether that principle must stand or fall, so I propose to test with ethical axioms the institutions as I find them, and thus determine their conflict or harmony with natural law.

The main problems of social science must be regarded as still unsolved. We know that results are unsatisfactory. We know that our most cherished institutions have failed to bring about the happiness which we all seek. The question for us to answer is, "Why?" Is our failure to accomplish our ends due to our ignoring natural law? Are our misery and our injustice, our wrongs and our affliction, the penalties for violating nature's decrees; or are the laws of nature essentially vicious and capable of producing evil results only? These are the questions which political and social science is called upon to solve.

I know that it is said that political economy is interested only in things as they are, that the science is practical and not speculative. What is knowledge for, anyway, if not for the use of man? If man seeks happiness why should he not use his reason-acquired knowledge to attain it? Is not that most practical which points out to man his errors of the past and helps him to build better in future? Why should we codify errors into laws and call the aggregation a science, while there are truths on which that science may rest? It does not make them the less errors that men have adopted and enthroned them. The history of the world is but a record of the errors and follies of mankind. Happy the nation without a history. Political science must learn a lesson from mathematical and physical science and, from failures and errors observed, learn truth and success. The history of sociology is no more the science of sociology than history of nations is the science of government. A catalogue of the principles used in trade may be useful in determining the principles of political economy, but they are not the principles of political economy.

CHAPTER II.

Man is humanity's greatest enemy.

In the world of industry there are disastrous phenomena which everybody has observed but which no one seems able to explain. These strangely occult manifestations have just recently been more apparent than ever before. The freest and richest country on earth has just experienced a business depression the most severe in its history. Banks collapsed, commercial houses closed and factories ceased to operate. Railroad equipments lay idle in the yards, steamships plied empty. Thousands of men, after stalking the streets of their homes in the vain search for employment, organized themselves into bands, overran the country and besieged the national capitol. Their cry was for bread, for themselves and their starving wives and children. The general government was compelled to coerce workingmen to yield to the demands of their employers. Police officers were, and still are, busy suppressing free speech and action lest it should lead to anarchy. He who expresses a new thought is tabooed as an enemy of society. The usual tax on the commmerce of the country is not sufficient to keep in motion the wheels of government; and the nation, after thirty years of unexampled peace and prosperity, is obliged to plunge further into debt for funds to carry on its ordinary functions. This is not a healthy picture. It must be a consequence of some principle at work in our institutions, yet no one seems to know what that principle is.

19

The reasons given for the strange situation are as numerous as the sources from which they come. Ask the politician, and he will say, perhaps:. "The use of silver as a money metal is the cause of hard times." The man at his elbow, who came from another section, will respond: "Not so; the threatened demonetization of silver is the element of mischief." "Good enough," says a third, "but the main cause of distress is the tariff." "I beg to differ," breaks in another, "the fear that the tariff will be meddled with is ruining the business of the country." "Want of confidence, that is our trouble," wisely remarks a philosopher. Confidence in what? But the meaningless platitude will not admit of specific statement. Another set of "thinkers" do not know; "such panics are necessary and their causes entirely occult."

The United States experienced a like panic in 1873, and the causes to which it was attributed were quite as vague and varied. The panic of 1857 was as marked, its cause quiet as indefinite. The panic of 1837 was caused by Jackson and his trouble with the banks, or almost any other cause one might name.

The panics of '48, '84, '64, '24, etc., were fully as inexplicable.

We may search the financial histories of countries with industrial systems similar to our own, and we shall find that in all crises occur with more or less regular periodicity. England has had them for hundreds of years.

There are those who see no menace in this periodic business prostration. They piously assert that it is natural and necessary and it is useless to fight against it. Any move toward change is an absurd attempt to make the world over again; any one who sees a wrong is a "calamity howler," whatever that may be. Although man has developed the present luscious apple from the sour and meager crab; although he has developed the ponderous draft horse and the lithe, supple racer from the little, long-haired, ungainly pony; although he has produced the massive Durham or Hereford from the little, nervous, long-horned ox of olden times, he is powerless to use his reason to better his own conditions. The beautiful watch-dog, the truest friend of man, has been bred from the treacherous, prowling wolf, by man's intelligence. Steam has been harnessed to do man's bidding and, by means of it, industry has been revolutionized; gardens have been made of deserts and deserts of gardens; the whole face of nature has been transformed by man's intelligence; yet we are told that using that intelligence to regulate the relations of man to man is absurd and foolish. It is, in the language of those philosopher-dilettanti, "really quite ridiculous" for human beings to use intelligence in ameliorating social conditions.

Such stuff as this would be considered too utterly nonsensical to require even a passing notice but for the fact that newspapers and periodicals of the highest standing lend their columns to the preaching of this doctrine of infantile social impotence.

Puerile aristocrats seem to forget that it is by human intelligence that all government has been established. If it were not for the use of more or less human intelligence in regulating the conditions of men, the physically strong would still be masters of the rest of mankind, and nine-tenths of the human race would still be in a state of chattel slavery. If it were not for a modicum of intelligence exerted in fixing the relations of man to man, neither property nor personal rights would have any security whatever. Our institutions may, to some extent, be growths, but yet they are crystallized by human intelligence from the mother liquor of crudity and barbarism.

The patrician apologist for a Chinese conservatism in treating things that are, would not be flattered to be called an anarchist, but he is more deserving of that title than the majority of persons to whom it is applied. He inveighs against the very foundation principles of government. If it is absurd and foolish to try to regulate the relations of man to man, then all government is absurd and foolish, for government is but an attempt to regulate the relations of man to man. If it is practicable to regulate these relations or even modify them

in one respect it is in all, and an attempt in one direction is no more absurd than one in another direction.
There is no generic difference in men associating themselves together for the purpose of securing to each his
life and property from the assaults of the armed bandit
than in associating themselves together to resist the
encroachments of the industrial bandit whose weapons
are wealth and unscrupulous cunning. Yet one is lauded
to the skies and the other is said to be absurd and
foolish. We have a school of *soi-disant* philosophers
who advocate anarchy in everything except the police
power of government, and why they do not in that
is not quite clear. Their physical courage is probably
lacking.

These same patricians and their apologists see no
wrong in overgrown fortunes, no menace to liberty. The
multi-millionaire is, they assert, an unmixed good without whom the poor would be infinitely worse off than
now. They do not condescend to give reasons for such
assertions, but state dogmatically that such is the fact.
Great inequality in wealth is desirable as well as advantageous. It is well that one man may be able, without the least inconvenience to himself, to smoke cigars
and drink wine to the value of several dollars each day,
although he renders no service whatsoever to any one;
while others, who toil unceasingly in the production of
wealth, must be content to live and rear a family on a
dollar or two per day. We might point out that this
is contrary to the maxim of justice: "To every one according to his works." We might show by hundreds of
instances that the man with millions is dangerous to
popular government, because he may influence elections
and legislation for selfish ends and pass laws contrary to
the will of the people. We might show that the vastly
wealthy may corrupt, degrade and oppress the citizen
and subvert the institutions of a great nation. We
might show that under the influence of enormously
wealthy individuals property might be held more sacred
than human life or liberty, and freedom might become
a hollow mockery. We might show that neither happiness nor material prosperity can long survive the fall

of individual liberty. We might show that the accumulation of large fortunes by the few at the expense of the many, has actually tended to produce the above results; but of course the plutocrat or his apologist could see no wrong in all this, and the effort would be useless. He is not even moved by the extreme degradation of abject poverty nor the menace to civilization of poverty-bred barbarism. He is too much wrapped up in selfishness to see or acknowledge anything; but fortunately for the country, there are not many so morally and intellectually obtuse as he.

The majority will not deny that a grave problem confronts the race, and that we are still far from its solution. Most men differ only on the method of solution, or on the question of whether the problem be capable of solution at all.

Never before in the history of the world have so many plans for the relief of humanity been brought forward as in the last twenty years. This economist sees in profit-sharing the full measure of human felicity. The remedy of that one is the full control by the state of all objects of monopoly. One asserts that the organization of laborers is the panacea for all social ills; another sees salvation in the education of the masses. Another believes in the single-tax, and another still in military socialism.

The industrial conditions of the immediate past, if not of the present, are more than a problem. They are the symptoms of social disease calling for prompt and effectual relief. The most dangerous manifestation of this ailment is the unequal distribution of wealth.

CHAPTER III.

We may divert the stream, but cannot stop it.

MEN toil as long and arduously as ever and their toil is far more productive, yet those who toil become none the richer. The more wealth produced, the more idlers there are to use it and the greater the number of people clamoring for bread. The more productive the toiler's work, the more extravagant become the lives of those who toil not. Ever is there found a way to divert this hard-earned wealth into the lap of luxurious ease. A woman who never produced a dollar's worth of wealth or anything else, will spend enough on one gown to keep half a dozen families of laborers for a year. Her husband or father or brother, or whoever she depends upon nominally, for support, if she does not levy her tribute directly, is as idle as herself.

The young millionaire who never accomplished a task useful to anybody except himself, if, indeed, to himself, spends enough on one debauch to educate and start in business the son of a productive toiler. What is wasted by the wealthy, in excess of that which is required to give them a good livelihood and gratify legitimate pleasures, would probably feed the hungry, clothe the ragged and shelter the homeless of the land. It would at least furnish a means of giving work to the idle and making them self-supporting. Why is it not used for these purposes? Where does the wealth that is so lavishly squandered by the rich come from?

It certainly does not make itself, as we shall see hereafter. It is evidently a portion of what numerous la-

borers produce, and the families of these men have not
at the same time enough to eat. We assert by our
laws that these luxurious idlers have the right to revel
in the laborers' wealth.

Why, then, are the masses so poor? Evidently because the classes are so rich. There is not wealth enough
to go around when so much is wasted. Where, then,
should intelligent beings look for the cause of poverty
and distress? Manifestly in the artificial economic
laws which derange distribution by allowing luxurious
idlers to take a portion of the laborers' toil.

If one-half the members of a family are spendthrifts,
it is easy to determine why the industry of the other
half will not thrive. Why does not the same rule apply to the great national family? If there was a rule by
which two brothers could take the bulk of the wealth
produced by the industry of two others, and live in ease
upon it while the two who toiled remained upon the
borderland of want, all could easily see the injustice of
the proceeding. But when the two parasitic brothers
increase to hundreds of thousands, and the toilers to
millions, we tacitly admit that the idlers of the family
have a right to the wealth the toilers produce. That the
wealth which they use is what the toilers produce, I
hope to show hereafter, as well as to point out the
method by which it is taken. Nobody who understands
the situation will have the hardihood to say that such
a proceeding is just either in the case of the four or the
millions; and whatever the edict of popular prejudice
and ignorance, aspiring teachers and philosophers
should not hug vain delusions. Nearly all tacitly admit
that the results of the distribution of wealth are not
what they should be. Wealth is distributed according
to fixed laws. There are certain rules as to what percentage of the results of production shall go to the toilers, and to the possessors of accumulated wealth. If
these rules were just, their results must be just. But
the result of these rules appears, to the thinker at least,
a monstrous injustice. The rules themselves must be
unjust. We have, to say the least, ground for suspect-
ing their injustice. The most important rules of dis-

tribution are the laws of rent and interest. They are
the basis of our economic system. They must be taken
before the bar of justice. The condition of the super-
structure demands it. The relations of the classes and
the masses, as well as the state of business, loudly cry
for a re-examination of the basic laws of our economic
system. The laws of rent have been examined by a
master whose work I shall not attempt to improve
upon.* I shall refer to them only sufficiently to make
other matters clear. I will examine the justice of interest-taking and its influence upon the distribution of
wealth.

Three elements enter into the production of wealth:
land, the laborer and capital, for which are paid re-
spectively rent, wages and interest, or capital profits.
Rent goes to the landlord, capital profits or interest
to the capitalist, and wages to the laborer. These three
get practically all of the gross product of industry. If
they receive their returns in just proportion, the inequality, the want, the misery of the earth are absolutely
necessary and it is foolish and vain to rail against these
conditions or ask for change. If the capitalist, laborer
and landlord, each gets his just proportion of the wealth
produced, the question is forever closed. We may plead
with him who gets the greater share to divide it with
his brothers, but we cannot ask him to do so on the
ground of justice, much less ask the government to coerce
him to do so. The problem, then, is to ascertain what
proportion of the wealth produced goes to the toiler as
such, the capitalist as such, and the landlord as such.
It is necessary to ascertain what each claims and what
each receives and what right he has to it. When that
shall have been fully accomplished, with how and why,
economic science will have completed its mission.

I use the term "laborer" in its broadest sense, including in that category all who toil with either hand or
brain to add to the gross assets of the human race;
but I rigidly exclude all whose efforts of either hand or
brain are directed to taking from others what others

*While there are many of the details of the theory of Henry George which I
cannot accept, I believe his main idea to be correct. I do not deem his remedy
quite effective.

produce. I exclude all of those employed in transferring the title of wealth from one to another without
giving an equivalent for the wealth transferred; rather
than in wringing from nature something which will add
to the wealth of all. He is not a productive toiler who
abandons the ranks of those who combat with nature
for sustenance and advancement, and skulks in the rear
of the army employing his brain and muscle to appropriate the spoils of the hard fought battle.

Wages, or the remuneration of labor in the sense in
which I use it, would include wages of superintendence,
wages of necessary financiering, wages of management;
in fact, all remuneration for human services which are
instrumental in adding to the sum total of the products
of the people's industry. I use the terms broadly in
this discussion, so that there may be no confusion arising from the fact that the capitalist is oftener a laborer
or a landlord also. What he receives as a toiler we
will charge to the account of wages, what he receives
as a capitalist to the account of interest or capital profits, and what he receives as landlord to the account of
rent.

CHAPTER IV.

" What thou wouldst highly, that wouldst thou holily,
Wouldst not play false, and yet wouldst wrongly win."

LEAVING for the present the question as to whether the toiler has a right to greater remuneration for his labor, let us inquire whether the remuneration which he does actually receive is too little or too much; whether it is desirable that he should receive more.

An examination of the average wages of toilers will convince one that those who do the work of the country are, at least, not overpaid. The census of 1890 gives about $445 per annum as the average wages of persons engaged as employés in the manufacturing industries. Experience teaches all of us that this estimate is very large. Any one who takes an interest in the subject can in any city find thousands of laborers working for less than a dollar per day, and wages over one and one-half dollars per day are the exception in occupations requiring little skill. But taking even the census figures as correct, the average given includes the wages of skilled, high-priced mechanics and foremen; hence some of the laborers must necessarily work for much less wages than the average given by the census compilers.

No one who has lived in a city of over fifty thousand inhabitants, the places where the bulk of our manufacturing interests are located, will assert that a family can be fed, clothed and educated as the citizens of a free nation should be, for four hundred and forty-five

28

dollars per annum.   Rent itself, or rent and street railway charges, will reduce the sum to three hundred dollars.   Even at this figure the family must live in the
cheapest and humblest of dwellings.

Seventy-five dollars per annum is the least possible
amount on which a family of four can be clothed. This
would leave but seventy-five cents per day for food, fuel
and incidental expenses, such as medicine and doctor's
bills.   Whatever the apostles of soup-bones and neck-
steak may figure out as the cost of living, any one who
has tried to live decently in a city on less than six hundred per year, has found himself confronted by a most
difficult problem.   The very fact that the greater number of bread and soup economists spend twenty times as
much on their own families as the amount which they
say is enough for the family of the laborer, would brand
their estimates as totally worthless.   Then if four hundred and forty-five dollars per annum is an inadequate
remuneration, one on which the toiler can scarcely exist, much less provide for the contingencies of the future, how do those live who get but one dollar or less
per day and that for a part of the year only?   This is
the condition of the common laborer.   I am sure he is
not overpaid.   If he has a right to more he has use for
it.

The census estimates of the wages of officers and
clerks of manufacturing corporations are quite worthless.   At least the average wages of this class gives no
idea of the wages of the extremes.   The average wages
are given at about eight-hundred and thirty dollars per
person per year, yet as this result averages together
thirty-dollar-per-month clerks and fifty-thousand-per-
year corporation officers, it gives no idea of what are
the wages of the greater number.   Our experience has
taught us that the average wages of the clerk is much
below the figures given and the average wages of the
officers or firm member almost infinitely above it. The
average clerk in our cities has a good deal of trouble
to make ends meet, while some corporation officers
are much overpaid.   This is especially true of railway
and bank officials in some localities, while in others

the incomes of these men come from interest and capital profits almost exclusively. Indeed where a corporation officer is paid over five thousand per year, his wages represent capital profits rather than services, for five thousand dollars per year will secure the best energy and mind of the country. But more of this hereafter.

There is no way to get at the exact wages of the farm laborer, but it is undoubtedly ridiculously small. About forty-three and one-half millions of persons live in the country, and nearly all depend for livings directly or indirectly on the products of the farm.* This product is but $2,460,000,000 per year. It pays charges other than wages, of at least half a billion. Now divide the remainder among the nine million or so of toilers who live in the country, and each is given the magnificent sum of $216 to live upon for a year and support the three or four others dependent upon him. If we include services in the estimate but about one-sixth of the wealth produced is produced on the farm and two-thirds of the population tries to live upon it.

Any one who within the last few years has seen life on the farm, with its unending, lonely, dismal round of drudgery from "sun to sun," with its pinching and shaving to make ends meet, while the mortgage ever gets a firmer grip (unless he be an Atkinson), needs no long drawn out argument to convince him that the toiler on the farm gets none too much for his labor. The honest observer cannot deny that the small farmer and the farm-laborer are being pushed by the grim might of poverty close to the danger line which separates the sturdy yeoman citizen from the ignorant, squalid peasant of the old world. If those who are inveighing

---

*In Census Bulletin 99 the following table of gainful occupations is given:

(1) Agriculture, fisheries and mining	9,013,201
(2) Professional service	944,323
(3) Domestic and personal service	4,360,506
(4) Trade and transportation	3,325,962
(5) Manufacturing and mechanical industries	5,091,699
Total	22,735,661

Portions of the second and third classes are dependent on agricultural product for a living, swelling the workers to be paid from agricultural returns a little short of ten millions, and making the whole population dependent thereon nearly up to the Census figure for country residents.

against foreign pauper labor would use some of their energies to save the American laborer from the same fate, they would be working more to the purpose.

We have thus far been speaking of the relatively prosperous laborer. We have not taken into account the sweater whose pittance is so inconsiderable that it is amazing that any one can live upon it. Then we have the underpaid factory girl, and the shop girl who, scarcely earning enough to keep body and soul together, is obliged very often to resort to other means to accomplish that end. When a girl is asked to work in the shops of our big cities at from three to six dollars per week, and is told by her employer or his agents how she may pleasantly supplement this pittance, it is not difficult to see why there is a social evil of ponderous proportions. Compelled to dress well to hold her position, unless she be fortunate enough to live with parents, she has the alternative of want or dishonor. Oh, ye luxurious, complacent matrons who wine and dine and sit on velvet cushions on Sundays, and listen to metaphysical discourses which aim at nearly nothing, are not you ashamed that your ease and complacency is often earned at the expense of the soul of your poor sisters, who if given but their just share of what you squander would be as honorable as you? Ye woman suffragists and W. C. T. U.'s, there is missionary work for you in your own churches among your own husbands, brothers and sisters. Teach the women who lead the society of our great, cold, extravagant cities to spurn the luxury which is wrung from the want of poor toiling fellow-women, teach them to honor men who would rather be just than to lie or cheat or oppress to gratify the vanity of the woman he loves, and you will have done more for humanity than you can do by a century of preaching virtue and practicing and applauding wrong and oppression. Men have done more wrong to gratify feminine vanity than from all other motives combined. One of the greatest wrongs of to-day is the starvation wages of the shop-girl.

All this in passing. It is evident that even those who

are not in need of work are underpaid for what they do. But this is not all.

It goes without saying that the industry of the nation should be sufficiently productive to support all of its inhabitants with reasonable comfort. Yet we have more than a million of laborers willing to work who are obliged to subsist on a pittance doled out by charity. Even of those who do work during their days of strength and vigor, one out of ten goes to a pauper's grave and two of the remaining nine are saved from the potter's field through the munificence of friends. At the wages received, it is utterly impossible for the common laborer to provide for his declining years.

From whatever standpoint we view it, we must decide that the remuneration of the laborer is too small rather than too great. This is the fact noted by all impartial observers. Every statesman preaches and every reformer asserts that higher wages are desirable.

Methods by which higher wages may be paid are as much sought as was ever the philosopher's stone.* Protectionists and free traders, gold men, silver men and greenbackers, populists, trust advocates and millionaires, all avow that what they do is to bring about higher wages. But higher wages can come only by trenching on the remuneration of the landlord or the capitalist or both, for they get all the product which does not go to the laborer. With the same amount of capital employed and the same amount of land and the same amount of labor and skill, other circumstances being equal, the product is constant. The only way to give more to one of the three classes among whom that product is divided, is to give less to one or both of the other classes. The important question to be solved is, has the laborer a right to more of the product and the landlord and capitalist to less?

*It is strange to see on one side of a newspaper a long article showing that $15,000 per year was not an extravagant allowance for a child not yet in its teens, and on the next page one showing that laborers are well paid at a dollar or so per day and that they have really nothing to complain of.

CHAPTER V.

THE MARGIN OF PROFITS—Some figures on the actual shares of the products of industry received by the several classes—Mr. Atkinson's cotton mill—The share of the laborer—The remainder goes to keep up wealth and remunerate the capitalist and landlord—Ten to fifteen per cent upon capital employed, the probable margin of profits—This includes interest and rent—It applies to successful enterprises only—-Census figures—Their indefiniteness—The margin small—To whom shall it go? the vital question.

"Our sins return to plague us."

It is extremely difficult to give precise figures as to the actual distribution of the gross products of the nation's industry, but from census figures and other sources this may be approximated. According to Mr. Atkinson's real or imaginary cotton mill, the laborers in a single process of production get 28 per cent of the gross product. This includes clerk hire and the labor of transportation. In the production of raw materials the laborer gets a somewhat larger share of the gross product, so that following an article through the successive stages of production from the time the raw materials are taken from mother earth to the time the product is given into the hands of the consumer, the laborer gets something more than fifty per cent of the total gross product. The rest of the product is absorbed by the landlord and capitalist and used in offsetting the natural deterioration and loss to which wealth is subject. A portion of the product received by the laborer is used in the payment of taxes, as is also a portion of that which falls to the share of the landlord and capitalist. Atkinson figures out that there is but six per cent left to the capitalist, or sixty thousand dollars on a' capital investment of a million, after all other charges are paid, and this, too, where business is done on a very large and economic scale. A portion of this is capital profit or interest and a portion land rent. This is net,

33

while the gross return to the laborer is given. If we first allow for charges for keeping up capital and insuring against loss, we should allow the laborer enough to keep up his strength and produce his kind as well as insure his life before we reckoned his wages. But even as a net profit this is on an average too small for the returns of successful business enterprises. The business man counts on his capital paying interest at about six per cent, as well as profits or dividends of about the same amount, which would make the net return to the capitalist and landlord about twelve per cent of the gross product of industry.

According to the figures of the census for 1890, but about ten per cent of the gross product was retained by the capitalist as profits and interest, making the return for the capital invested about fourteen per cent. This is not exact, as capitalist and landlord profits enter to some extent into the item of miscellaneous expenses as well as that of salaries of officers and members of firms.

Then, in the census estimates, no allowance is made for the charge for deterioration of capital, which would amount to two and one-half or three per cent of the gross product.*

All this goes to show that the margin of profits is not large, and the vital question is, to whom do these profits belong? If they should go to increase the wages of the laborer, they would be found sufficient to give employment to the involuntarily idle and save those who do work from the pinch of poverty. If they rightly go to the support of the idle rich there is no cure for poverty and want. Of course with more productive workers, more would be produced.

*In the summary prefacing extra Census Bulletin No. 67 we have the following statement of expenses of manufacturing: Miscellaneous expenses, $630,944,058; total wages, $2,282,823,265, total wages (officers, firm members and clerks), $391,914,518; *all other employes*, total wages, $1,890,908,747; cost of material, $5,158-868,353. The total value of products is given at $9,370,107,724.

CHAPTER VI.

THE LANDLORD—His right to remuneration—Land defined—Ricardo's definition—The significance in which it is here used—Mines not land in the same sense as farming land or building sites—Forests not land, unless cultivated—The natural basis of property—What one produces is primarily his own—What he does not produce or what another produces is not primarily his own—Land not produced by individuals—Does not belong to individuals—Belongs to the community in usufruct only—Community can give no greater title than it possesses—The collection of rent does not aid productivity—No return given by landlords for rent taken—An imposition—Royalties in the same category—Landlords and royalty-takers levy a billion of dollars or more per year on industry—They take that much of the annual product—It does not justly belong to them, must therefore belong to laborer or capitalist or both.

"What are we otherwise here than guests?"

WHAT right has the landlord, then, to remuneration? Land, in the broader signification of the term, is the free gift of nature to all her children. It is the coal in the mine, the lead, iron, gold and silver in the ore, the standing forest, the fish in the stream, the water to run our mill-wheels and float our fleets, the firm earth on which to build, the air, the imperishable qualities of the soil. By Ricardo the term land is limited to the imperishable qualities of the soil, and this interpreted to include streams and water powers is the significance in which I shall use the term. It is for land thus limited, or rather the values attaching thereto, that rent proper is charged. Ore, coal, timber, fish, etc., in their original places, or the mines containing them, while supplied by nature are not land in the sense that fields and building sites and roads and navigable rivers are land. They are perishable, limited in quantity, and are consumed by use. The compensation charged for them is not rent proper, and does not follow the laws of rent. To secure the value of the products of mine and forest to the whole community, will require very different remedies from those required for imperishable land.

Land is the undiminishing contribution of nature to the needs of man. It is given value by the multitudinous improvements and adaptations civilization developed upon it by the heads and hands of toil, taken in connection with the fact that it is limited in quantity, requisite as a basis for all production, and attaches to itself the utilities of a portion of the improvements upon its surface. Mines and forests are natural storehouses not spontaneously replenished by nature; land, what is left after the store has been taken.

It will be readily admitted that what one produces is primarily his own. This is a necessary corollary to the equal rights of all to life, a right which is affirmed by the practice of all civilized communities. It follows, necessarily, that what one does not produce is not primarily his own. These propositions are the basis of all property. If one can not gain a property right by production, there is no way in which he can gain such a right. And everything which man uses, except what nature freely gives, is produced by man. One can gain title to what he has not produced by a free gift from the producer, or by trading directly or indirectly what one has produced for it. A title to what one has not produced can come only through him who has produced it. Applying this axiomatic principle, the landlord has no just title to land, no greater right to it than any other citizen. He did not produce it, for by definition, it is the free gift of nature. He did not gain title to it from him who produced it, for no human being produced the land, and nowhere is it written in the record of nature that any individual shall have exclusive title deeds to that which is the common inheritance of all. Even the community can give no greater title to land than they themselves hold, and this is but usufruct for the passing generation only; and then only on such terms that the thousands born daily into the world shall share the gift of nature as well as those already here. For these new-comers have an equal right to live with those here before them, and none can live except by utilizing nature's gifts. Nature has no laws of primogeniture. Her bequests are to all the generations of men.

As the landlord has no just exclusive right to the land, he has no right to charge others for the use of it, and no right to withhold it from the use of others. He

may have a right in usufruct while he lives, just as has
every other individual, but this is only to what he can
personally use to ordinarily good advantage. What he
does personally use must be the criterion of what he can
use. Rent belongs to the community as a whole. For
upon the improvements built up by the community as
a whole the rental value of land depends.*

The fact that the landlord's collection of rent, or the
owner's natural advantage, his royalty, makes the farm
no more productive, puts no more gold or iron in the
ore, no more coal in the mine, makes the site no better
to build on, the river no more navigable, the water no
richer in fish, the forest no more extensive, the stream
no more powerful, stamps the rent and royalty charge
as a tribute levied without any return. These charges
are founded solely on cunning dishonesty and force.
They have not the shadow of an ethical sanction. Rent-
taking by landlords, then, is not only grossly unjust, but
is in no way necessary to production. It is a charge on
production for which there is absolutely no return. It
can then be well dispensed with. By all the canons
of business, when one is making retrenchments the first
expenses to attack are those which are unnecessary.
But these are worse. They are absolutely inimical to
the best results in production. Land is the basis of all
production, and the practice of allowing the landlord
to collect rent on land has led to a monopolization of
the earth which deprives vast numbers of willing producers of an indispensable instrument of industry.

What is true of the landlord in this respect is true of
the exacter of royalties also. Besides giving absolutely
no return, he discourages and cripples industry.

According to a bulletin of the census of 1890 the real
estate of the country, lands and improvements, was
worth in that year thirty-nine and one-half billions in
round numbers This would place the value of bare
land at something less than twenty billions. Rents are
not more than five per cent of this sum, giving a yearly
aggregate of one billion paid to the landlords of the

*That this proposition does not express the whole truth will be shown hereafter.

country. Profits on land held speculatively must be added to this, swelling the aggregate very materially.

It is impossible accurately to estimate royalties from forests and mines, but they aggregate hundreds of millions.

We have then considerably more than a billion dollars per year exacted from industry without a shadow of right or a pretense of valuable return. Neither the landlord nor the appropriator of others of nature's gifts has any claim to it in justice. It must belong to either the capitalist or the laborer. Let us inquire which of these classes have the better right to the money now paid in rent and royalties. In doing this we will also put the test of ethics to the share which now goes to the capitalist.

The single tax proposed by Henry George and his followers, if capable of application, would greatly mitigate, if it would not destroy the evil of private appropriation of rents. It is probable that nothing short of state ownership of mines and close state control of forests can ever give royalties to the community, the only just owner. It looks to me as though a tax on mining land would scarcely meet the requirement, for the tax must be levied in accordance with the output, especially if it were intended to take the whole royalty charge, and then it would become a tax on the product capable of being readily shifted upon the consumer. But more of this hereafter. In fact, there is no such thing, economically speaking, as mining land. We have no soil capable of reproducing minerals, even by the application of any amount of labor.

CHAPTER VII.

Ambiguity is the stumbling-block of logic.

AT this point it is necessary that terms be clearly defined, for from the use of ambiguous terms arises nine-tenths of the ordinary confusion of thought. Capital is wealth used in production. Wealth is anything having value in exchange; i. e., capable of commanding a price. Buildings, all improvements on land and all movables are included in the term "wealth." The capitalist is he who controls wealth, and from that fact, apart from any individual services, claims a remuneration for allowing his wealth to be used. This remuneration is called interest or capital profits, depending on whether the wealth is loaned to another or managed by another under the direct control of the owner. In this inquiry it is unnecessary to distinguish between interest and capital-profits, for both are returns for the use of wealth, and not for the individual services of the capitalist. It is the capitalist, and not capital, who receives these returns; and it is the laborer, and not labor, who produces wealth and receives wages. We are inquiring into the rights of the laborer and the capitalist as such, and not into the rights of labor and capital, for the latter terms represent things without any rights whatever. I here use the term laborer in the sense in which I have used it before in these pages; he who uses his brain or hand in adding to the sum total of human assets, excluding him who uses his powers to appropriate what has already been produced.

39

These distinctions may appear pedantic, but they are absolutely necessary to meet a school of apologists who are either so mentally confused or so sophistical, that they readily accept conclusions based on using terms in different meanings in different portions of a discourse. Some of them go so far as to claim a remuneration for capital on the ground that it is labor.

It matters not so much whether terms are used strictly in their ordinary meaning, if the same meaning is attached to them in all portions of the same discussion. To show the vital necessity for a precise use of terms, I will notice briefly certain arguments based on an ambiguity of words and thus clear the way for subsequent discussion.

Capital is stored labor, say the apologists, and from this premise they go on to argue that since it is stored labor and was produced by labor, it has the rights of the laborer. There are several fallacies buried in this nebulous maze of hazy cogitation.

It does not follow that because a thing was produced by labor it is labor. In fact, that is the best argument that it is not. A force and the result of a force are quite different concepts. You may as well say that a spade is a man's ingenuity because a man's ingenuity produced it, as to assert that the result of labor is labor because it is the result of labor. The wealth resulting from labor, or the stored labor, as it is termed, can no more produce what labor can, than the spade can produce another spade. The reason in both cases is identical. The spade, though produced by man's ingenuity, is not man's ingenuity; and wealth, although the result of labor, is not labor. Wealth and labor have no more in common than a man's ingenuity and a spade. Wealth is an inanimate, decaying thing, labor a living force.

The confounding of cause and effect which pronounces wealth used as capital essentially the same as labor, is so utterly absurd that it would not be worthy of notice were it not for the fact that acute thinkers in other fields cling to this fallacy.

But these apologists go a step further in their argument. They say that the result of labor is labor, but the result of labor is also capital. Capital is therefore labor and hence has the rights of labor. Labor produces all wealth, hence the labor which is the result of labor, and is at the same time capital, is entitled to the share which it produces. Thus playing on different meanings of the terms labor and capital, and using these rigidly distinct concepts as interchangeable, these muddle-brained philosophers prove to their own satisfaction that capital is productive, and hence entitled to a portion of that which is produced by the laborer.

This is exactly the same sort of logic by which the tyro in that science conclusively proves that a cat has ten tails or that one cent is better than heaven. It is labeled by logicians, the fallacy of ambiguous middle.

It is scarcely necessary to point out to the man of common sense that what the laborer produces is not labor, no more than the spade is a man's ingenuity, or the bedstead which he produces is the cabinet maker. If you apply the term labor to these results of the toil of the laborer, you mean something altogether different from what you mean by the term labor used in the sense of the productive activity exerted by the laborer. What is true of the latter is not in any sense true of the former. If you change the term to stored labor and then to capital and then substitute the latter term for labor where it occurs, you are simply juggling with terms, and the results obtained mean absolutely nothing. Your action would be the same as though you solved a problem in mathematics by letting "X" represent thirty in one portion of the process and one thousand in another and substituting these different "X's" for each other—as though they were the same.*

If these philosophers had not made this mistake in the indiscriminate use of labor and capital, there would be another fatal difficulty with their argument. However, much the same fallacy would be involved. Neither capital nor labor has any rights whatever. They are simple things, with no more rights than a doorpost or a front fence. This is especially true of capital, which

*There can certainly be no fault found with using words figuratively in discourse, but it is an unpardonable fallacy to argue from both real and figurative meanings in the same discourse.

is nothing more than wealth intended for a certain purpose. Labor is a thing when used as a general term, but cannot be separated from the laborer. The rights belong to the laborer and the capitalist, and this brings us again to the original proposition lying at the foundation of all property rights. Man has a primary right to what he produces, and hence has not a primary right to what any one else produces.

The capitalist bases his claim to what is produced on the claim that his wealth produced it. The laborer bases his claim to what is produced on the ground that his labor produced it. This must necessarily be the basis of the claim of each, otherwise our first proposition would be wrong.

The capitalist, as such, certainly does not directly aid in the production of anything; for as a capitalist he toils not in production. As he has no title to what the laborer produces and can acquire none through his wealth, he has, as a capitalist, no title to anything produced, unless his wealth is capable of producing it. The whole question of the right of the capitalist as such to interest or capital profits, to any form of remuneration whatever, for allowing his wealth to be used, depends on whether wealth is of itself productive. It is a question of fact, purely, capable of being established by evidence. Let us put human experience on the stand and ascertain the truth.

The idea that the mere possession of wealth, regardless of whether that wealth be capable of production, gives a right to a portion of the product of industry is an obvious absurdity. The capitalist has not as a man a right to what the laborer produces. If he has any greater rights as a capitalist than as a man he must have gotten those rights from his wealth, for the possession of wealth is the only difference between the capitalist, as a capitalist, and the capitalist as a man. If he got from wealth the right to appropriate to his own use the wealth produced by the laborer, the wealth of the capitalist must have had that right. This would be asserting that wealth, an inanimate thing, had greater rights than either the capitalist or the laborer, which is manifestly absurd. Unless wealth is essentially productive, the capitalist, as a capitalist, has no right to any remuneration whatever. Is wealth essentially productive?

CHAPTER VIII.

"The proudest work of man as certainly, but slower,
Must pass like the grass 'neath the sharp scythe of the
mower."

Is wealth essentially productive?

Every article of wealth produced by man has within it the essential principle of decay and final complete destruction. Nature lends it to him but for a time; after a time she reclaims it as her own. The condition of the loan is constant use. Man must produce unceasingly to keep his stock of wealth intact. There are no exceptions to the rule. The more indispensable an article to humanity, the more prompt and certain its decay.

The most stable of man's works are the least useful. The vast pyramids of Egypt, useless monuments to the superstition and misguided ingenuity of a race of slaves, seem, at first glance, eternal; but, although their existence has covered but a point in the existence of short-lived man, the hand of time is already grinding them to the dust. Eternal Rome is in ruins; the palaces of the Cæsars have crumbled to decay. More terrible than the Goths and Vandals has been the edict of Nature reclaiming her own from the evanescent imprint of the feeble hand of man. Palmyra and Thebes are but half forgotten names; Babylon, but a symbol of iniquity. Scarcely less perishable than man himself are the works of his hands. Remove the preserving

44

care of the laborer from man-made wealth, and its destruction is but a question of days.

So true is this that we might conceive of a great dynasty of kings owning the earth and all of its bright cities and all of its teeming wealth; yet, if no toiler's hand were raised to save, the scions of that dynasty would starve as they watched their fair cities crumble, and the earth become a wilderness. Even after a quarter of a century, there would not be a king left to tell the tale.

If we turn our attention to articles of common use, we find them more perishable still. The staunchest ship will scarcely brave the storms of half a century. Place her idle and unattended in the docks and she will rot in a decade. The locomotive with its frame of steel and its coat of imperishable brass, if active, will scarcely outlive the youth of the hand that fashioned it; idleness will not extend its career. The average useful life of a machine is estimated at twenty-two years, and the rust of idleness will destroy it more quickly than the wear of work. What would become of our electric systems, the metallic nerves of mother earth, if abandoned to the destructive powers of nature for even ten years? We could scarcely determine that they had ever been. If abandoned for a quarter of a century, the continent would turn into a wilderness scarcely less desolate than when Columbus landed here. Our roads and streets and wharfs and shops, if left to themselves, would scarcely survive the hands that built them. Rats would gnaw where silk-robed judges sit, and serpents hiss where social revelry now resounds.

Consider the things most necessary to man; that which he eats and drinks and that with which he clothes himself. Let the laborer drop his hands, abandon elevators, cribs, storehouses, stables and herds to the worm, rat and weevil, to the inclement elements and the deserted fields, and humanity would be starving within a year. The earth would be a savage-populated wilderness within ten years In the matter of food and clothing, humanity literally lives from hand to mouth.

Why then this idle boast of the power and independence, of the productiveness of capital? Why fly in the face of facts and affirm that the capitalist can afford to rest and feed on what he has? If young Gould, the inheritor of his father's millions, refused, for a single month, to work with his hands, and others refused to labor for him, he would be in a worse condition at the end of that time than the meanest denizen of Whitechapel. If laborers deserted him to-day, not all the efforts of his puny hands could save even a wreck of his mighty fortune from the destroying hand of nature. He would be poor as a savage before he had time to turn gray.

Man-created wealth is not productive. The principle within it is decay, not growth. And bonds cannot save it, the edicts of capitalists can not save it—it is labor with the head and hands, and that alone which must and does preserve it. Humanity lives on man-created wealth. The imprint of the laborer's hand must be placed upon the treasures of Mother Earth before they become current in nature's great banking house. There are no exceptions to the rule.

The above are but a few examples of the inherent decay essentially embodied in all wealth, but the principle needs no other proof than experience common to all. Not a single instance can be cited of the spontaneous increase of man-created wealth; not a single instance, of aught except man-created wealth supplying the wants of man. The decaying quality of wealth is self-evident when thought upon.

But what is the assumption of the capitalist? How does he justify interest-taking? As we have seen, interest or capital profits are not a remuneration for the toil of the capitalist, for the capitalist as such toils not. As we have also seen, he has no claim on that which is produced by another, for that would be denying the right of every man to what he produces by his toil, and would overthrow all right of property. Then he depends for his interest and profits on the assumed power of wealth to produce more wealth. He must justify the taking of a return more than the capital lent on

the assumption that the wealth lent increased of itself,
for if it was another's toil which increased it, in taking
the excess he would be taking what another produced,
and therefore, what belongs to another.

But we have seen that the assumption that wealth
has within it the power of increase is contrary to the
facts in the case. That, without exception, the inherent principle in wealth is decay, not growth.

The practice of interest-taking flies in the face of
facts and asserts the producing power of unaided wealth
at every turn. This assumption, if carried to its logical
conclusion, would lead to very strange results. If interest-taking is right, compound interest-taking is right.
The principle of compound interest is, that a dollar, or
the wealth represented by it, without any exertion on
the owner's part will grow into two dollars in a given
number of years, four dollars in twice that period,
eight dollars in three times the original period, and
that it will keep on increasing in a geometrical ratio
until that one dollar, with its interest, would, in time,
represent all of the wealth on the earth. The rate makes
no difference as to the principle of the thing. Money
at compound interest will increase indefinitely at five
as well as at twenty-five per cent, though more slowly,
to be sure.

It does not require a philosopher to see the absurdity
of a principle deduced from the power of wealth unaided to increase indefinitely, yet our whole financial
system is based on this very absurdity. And what makes
matters worse, it is not one dollar that is assumed to
have the power of indefinitely increasing, but several
billions of dollars.

A syndicate of less than one hundred American capitalists, if allowed to collect interest on their capital
at a low rate and re-invest for 150 years or less, would
at the end of that time own the earth and all real and
personal property thereon. This is a simple mathematical proposition, capable of exact demonstration, and
any one who doubts the truth of this statement may
set all doubts at rest by computing compound interest
on one and one-half billions of dollars for one hundred

and fifty years, at five per cent per annum. Great corporations tend at present to extend their investments
and to decrease the number of their important share-
holders. Corporations live for centuries. A corporation coming to practically own the earth under the laws
of interest, is not only not impossible, but not improb-
able. We have already instances of corporations holding controlling interests in towns and cities and states.
One two-hundred-and-fiftieth of the population have,
under such methods, come to own eighty per cent of the
wealth of the country. Figures deduced from the census
have been so averaged and manipulated by writers as
to make it appear that nine per cent of the people of
the nation owned in 1890 but about seventy-one per cent
of the wealth, but an analysis of the figures will bear out
my statement. Either is startling enough. The only
difficulty in the way of a private corporation monopo-
lizing all wealth, seems to be in getting an organization
large enough, and we are rapidly overcoming that difficulty. Will any thoughtful man knowingly support a
principle which might give to one- hundred irresponsible brigands all the wealth of the earth, to the exclusion
of the other billion and a half of humans? The principle of interest-taking will do this. Its philosophy
is the acme of absurdity, yet all men seem to acquiesce
in the practice.

The more wealth is saved, the more there is to bear
interest. Hence, burdens forever increase. The wealth
of the world is an inverted pyramid, the misplaced base
of which becomes more unwieldy day by day. The
interest-bearing wealth increases in a ratio which is
ever growing more and more rapid. It is a very well
established fact, or rather law of economics, that the
power of producing wealth decreases in reference to the
labor expended, after a certain limit is reached. This
law applies to the bulk of wealth. It is called the law
of diminishing returns.

To be more specific: After a certain fixed limit has
been reached, the return which land yields to the application of additional labor, is comparatively less. All
wealth is produced by the application of labor to land.

We have, then, under the law of interest, liabilities more and more rapidly increasing and assets growing proportionately less. The inverted pyramid becomes more and more unstable. No thinker worthy of the name can uphold a law which implies a flat contradiction of the precepts of nature, and whose logical consequence is to keep the world forever tottering on the brink of bankruptcy.

Wealth cannot be produced with sufficient rapidity to keep pace with the demands of interest. The loaned capital must necessarily absorb all of the wealth and the money lender become possessed of all of the property on earth. Land is subject to private ownership and may also become the property of the money lender. The laborer then will be absolutely at the mercy of the capitalist. Deprived of land in his own right, he must use the land of another. Deprived of capital in his own right, he must use the wealth of another. All that he produces, more than is barely sufficient to keep him alive, must go to the capitalist in interest and to the landlord in rents. He must take the terms offered him and live on what he is allowed by his masters. If these masters do not wish him to live at all, all that is left to him is to break the law or die. The undertaking business man must use the wealth and land of the capitalist and landlord, or collect rent and interest for his own, in addition to remuneration for his toil. He also is accustomed to set apart a percentage for profits. If any capital used in business collects interest, all capital used in business must tend to collect interest. For if a business man could command as large an income by loaning his capital and avoiding the risks of business, as he could by engaging in active business, he would lend his capital and leave active business alone. His object in becoming an active business man, is to gain both profit and interest, often in addition to wages for his toil. His venture fails of his object if he does not succeed in gaining both. This will sustain my large estimate of interest-bearing wealth.

If the business man employs laborers, those laborers are obliged to produce the wealth which is given in interest. If he be simply a laborer, employing his own capital, as are so many small farmers and tradesmen, he must make his labor produce interest and wages or, in comparison with the money lender, lose either time or interest. Only capital allowed to lie idle or which is dissipated in unfruitful undertakings, fails to exact interest, and the latter is totally lost, while the former is a charge on the whole community.

All our energies are bent toward applying the principle of interest. We undertake to pay increase on an enormous sum. According to the United States Census of 1890, the wealth of the United States, on a gold basis, was over sixty-five billions. Allowing for gold appreciation, it would foot up something like eighty-five billions, and we attempt to pay interest and rent on more than one-half of this enormous sum. If interest on a billion and one-half would in a little more than a century absorb all the wealth on earth, what will be the consequences of trying to pay interest and rent charges on thirty times that sum? This is why the interest question has become one of vital importance. It is the great question of the day.

CHAPTER IX.

THE PRODUCTIVE POWER OF WEALTH—Instances and objections—Is capital productive?—How it differs from wealth—The productive power of a machine —Interest not a charge for a machine's effectiveness—Interest a charge for use of wealth—Is the borrower alone benefited by that use?—Men should be paid for what they relinquish, not for what they can extort—A scientific test of the productive power of wealth—How we determine the cause of steam—The verdict against the productivity of wealth—Wealth an advantage to the laborer, but that is not the question—The toiler is heir to all the ages, has a right to the use of ideas of inventors applied through wealth -Unfinished product—Other instances of decay not growth in wealth—The house—The mill—The horses.

"WEALTH cannot produce, but capital can," wisely remarks the apologist. "Wealth used in production is productive. Does not the locomotive do more than a thousand men?" Even if that were true, it would in no way affect my argument, for there is but one class in the world which can convert wealth into capital. That class is made up of those who toil. While wealth remains in the hands of the capitalist, it is simply wealth. When he lends it it is wealth, and he claims interest for it whether it is scattered to the four winds of heaven or used in productive business. It is only when the life throb of the laborer pulsates through the decaying form of wealth that it is animated into capital. It is this life-throb which wrings additional wealth from reluctant nature. If capital is productive and wealth is not, then the laborer alone can make wealth capital, can make it productive, and he alone would be entitled to the resulting product

But capital is not productive. It is wealth and nothing more, having no inherent quality different from other wealth. It is known by a different name because it is being used by a productive toiler. It is a tool in the hands of the laborer, a tool which rusts and rots and wears out just like other tools. As a tool, the laborer uses it to make his labor more effective, but the productive power is in the laborer and not in the tool.

51

Human energy aided by human intelligence is the only
force yet discovered which can fashion nature's product
into forms which supply the wants of man. However
far removed from the result produced, it is human intelligence which produces the result. Just as the electric current and not the wire which transmits it moves
the needle thousands of miles away, so the energy of
the laborer, through the lifeless, decaying tool, shapes
the raw material, the gift of nature, into forms to satisfy the wants of man. Capital is at best but a medium
for applying human energy to production, not a productive element. You may as well say that it was the
pencil and paper which produced the poem, or the spade
which dug the potatoes from the ground, as that capital produced this or that quantity of wealth. The glowing spark of human energy, whether of brain or hand,
is the sole productive spark outside of nature's forces.

This truth is easily verified. A machine is one of the
most important forms of capital. The machine is made
up of three elements: perishable wealth, the idea of the
inventor and the labor energy of the toiler. The productive element in the machine, as a machine, is the
idea of the inventor applied by the energy of the laborer.
The wealth is the same dead, decaying thing, incapable
of even preserving itself. The locomotive is capable of
doing what a thousand men could not do without a locomotive. Is it the wealth in the locomotive which
does the work? Not at all, it is no more potent than a
pile of old iron. The inventor's idea, which planned the
way in which this wealth may be fashioned to be used
as an instrument, and the mechanic's skill which carried out the inventor's idea, and guided the wealth in
the application of that idea, are the creative elements.
Without both, the locomotive would be a mass of dead
wealth and no more.

The capitalist, as such, is not an inventor; he toils
not with hand nor brain to add to the wealth of the
world. He is not a laborer. Then the capitalist who
owns the locomotive can claim a remuneration for its
use, when lent to another, only from the fact that it is
wealth and not because it is a machine. He cannot

claim it on that account, for wealth is not productive, but is a charge on those who use it. Indeed the capitalist does not lend the locomotive or any other machine, as a machine, but as a specific amount of wealth. Unless he had an absolute monopoly on locomotives, the price which he could charge for one, or its value as interest-bearing capital would not depend upon its effectiveness as an implement of production, but upon the cost of its construction. For unless monopolized, the amount of wealth consumed in its construction would command such a machine, no matter how effective that machine is in production. But it is not claimed that interest is based on monopolistic powers; then it must be based on a claim for the use of wealth on the ground that such wealth is productive. This we have proved to be a false principle. If based on monopolistic powers this fact would give it no ethical sanction.

In determining scientifically the cause of any given phenomenon, when several observed elements are present in any given case, we eliminate those elements showing no positive tendency to produce the observed phenomenon, and look upon that element as a cause without which the phenomenon cannot be produced. For instance, in determining the element capable of converting water into steam. We conclude that the stove is not the cause of the water's conversion into steam, for the stove without the fire has no tendency to produce such an effect, while fire on the earth or in a grate will produce such an effect. In like manner we may determine that the kettle is not the cause of the phenomenon, and following that course we may eliminate everything but heat. Heat is always present when water is converted into steam, and when applied to water has always a tendency to produce such a result. We therefore conclude that heat is the element which converts water into steam. The stove, boiler, etc., may be useful accessories, but are not the causes.

Applying a similar test to the cause of the phenomenon of production, we are forced to conclude that the laborer is the sole cause of the phenomenon. A spade may lie away until it rots, and still no sod would it

upturn unless the strong hand of the laborer was upon
it. An ax would never fell a forest oak, unless there
was behind it the sturdy brawn of toil. A reaper would
never cut a harvest, without the guiding, sunburnt
hand. A locomotive would never haul a car without
the grimy fingers at the throttle. A needle would never
transmit a signal, without the deft ear and hand of the
operator to receive and record it. Yet the laborer placed
on virgin earth, without a vestige of man-made wealth,
upturned the sod, felled the oak, cut the harvest, trans-
ported wealth, transmitted signals and, stripped of all
man-made wealth, could do so again. Then the laborer
is the cause of production; wealth is not, capital is not.
That is the conclusion of scientific reasoning as well as
the verdict of fact. There is not a fact in natural science more conclusively proven than this.

But it is said that the wealth loaned by the capitalist aids the man who uses it, and that he should therefore pay for its use. Its being used aids the capitalist
far more, even though he never received a cent in interest for its use. The laborer who uses capital more than
repays its owner by keeping it intact. Nature in her
divine wisdom has decreed that wealth shall not be
hoarded. If not used by the hands of labor in producing more wealth, nature, after a few short years, reclaims it as her own. Does not the laborer, then, do
the capitalist the greatest of services by taking his
wealth, preserving it from the wrecking hand of time
and returning it to him intact? It is not an answer to
say that the laborer is at the same time producing more
wealth, a portion of which is for himself. By that very
act he keeps the world moving, keeps up the march of
civilization, keeps all of us from the fate of poverty-
stricken savages. Here again we meet with nature's
inexorable law. Toil or perish, is the decree pronounced
against the race. It is only by fraud committed against
the many that a few are exempt.

The laborer unaided has gained a livelihood. He
might do so again. For unaided capital, there is but
death and decay. How fortunate for the capitalist that
he can make the laborer his mediator! For there is

not one article of wealth which can survive without such mediation.

It is true that this does not agree with the tiger philosophy of early economists. Their motto was that one is entitled to all that he can get by any means by which he can get it; if, indeed, they troubled themselves little about rights. On this principle the capitalist can, by his advantageous position in controlling the implements of toil, wring interest from the laborer. But does that make it right? That is the principle practiced by the tiger in springing upon the deer. That is the principle practiced by the highwayman in holding up his victim. That is the principle which says rapine is just and honorable. Why not be consistent and allow the highwayman to take what he can, if ability to appropriate carries with it the right? The ethics of civilization, on the other hand, teaches: to every man according to his labor. Man should be paid for what he relinquishes in serving another, for the trouble and exertion it cost him. There is no other subjective criterion. Where benefits relinquished can be readily estimated, they are the sole ethical guide to deserved remuneration for any service. One knows what a particular act cost him, but not what amount of benefit it was to another. On this ground the capitalist would not only be entitled to no remuneration for lending, but would by that very act receive a benefit for which he should pay. And we must introduce this principle of justice into our sociology before we can expect to arrive at just results. Where ability to exact is the measure of remuneration, right is might, and might is force or cunning. Such a theory put into practice leaves upon avarice or arbitrary power no ethical check whatever.

To be sure the effort of the laborer, with the appliances which toiling hands and brains have conceived, fashioned and perfected through the ages, is more effective than is the effort of the naked savage. But this is not the question. The generation of to-day is the heir of all the ages. The inventive toil and ingenuity of all the past belong to all the people, and is more justly the property of him who uses them to advance the interests of the whole people, than of him who wishes to use them for extortion only. The toiler is the heir to the ingenuity of the toiler. If the capitalist has succeeded in taking the wealth of the toiler of the past, that fact gives him a claim to perishable wealth alone. It gives him no monopoly of the great inventions which brain and hand have wrought. And these inventions, and not the fact that they comprise a certain amount of wealth, give every implement or machine from the soup kettle to the printing press its value in production. Hoarded wealth, untouched by the hand of the laborer or the inventor, would be absolutely worthless in production. It would have less effect in production than the kettle has in producing steam. Look at it as you will, wealth is not productive, but the toil of the laborer is.

If making wealth capital does not relieve it from its natural condition of decay; if it is still a charge on him who uses it, and useful only after the toil of the laborer adapts it to the ends for which he intends it; if it is useful then only as a medium of applying the productive forces of the laborer, the capitalist has no claim for remuneration for the services which his wealth renders in production. It has potential serviceability only and that serviceability is incapable of being developed by any except the toiler. While the wealth remains in the hands of the capitalist, that potential serviceability is but a principle of decay. The laborer who resists that decay and developes that potential serviceability does the capitalist a substantial favor poorly repaid by the loaning of capital on the condition that it shall be returned intact. The laborer is not called upon in justice to give up any of the gross product of his toil, except enough to keep the capital which he uses intact and to support the state. There is no hardship to him in either of these conditions. The state affords protection for his toil, and keeping capital intact redounds to his advantage as well as to the advantage of all, for if capital were allowed to deteriorate, productive labor could be less effectively employed and the product would be smaller.

Even looking upon capital as unfinished product, the same rule holds good. Unfinished product from its very name is seen to be unfitted to supply the wants of man, the aim of all production. An unfinished product, like all other wealth, will become less and less valuable by the process of time. It will never finish itself, but waste away. The laborer alone can finish it. His finishing it adds to it all the value of the finished over the unfinished product, and he should have all of the difference as a remuneration, giving to the capitalist or the owner of the unfinished product all that the latter contributed to the final product, i. e., the value of the unfinished product.

Let us suppose that a man has a stable of horses which he cannot personally use, and the value of which he wishes to preserve for some future time. Would not a toiler be doing that man a marked service by taking these horses and using them for ten years, and at the end of that time returning to the owner an equal number of good young horses in their stead? This would be wealth lent without interest. (We are dealing with wealth, not money.) If the capitalist had kept these horses, they would, within the ten years, have all grown old and unserviceable, and he would, in the meantime, have had to pay for their keeping. Under the interest system, he would have compelled the toiler who borrowed his horses, not only to pay for their keeping, but, when the horses had grown old, to return to him two good young horses for each one taken. It does not require a philosopher to decide who, in this case, has the better of the bargain. We must keep the fact constantly in mind that the horses represent wealth which the owner cannot personally use at the time when he decides to lend it, but which he wants to preserve for some future time. Under the present system, he would sell his horses and put the money at interest; for, although horses become useless by the lapse of time, we have a fiction that the scraps of paper or bits of metal representing their value increase in worth with each rising sun.

There is a house on a principal street of a growing

city. The location is the best. The appointments of
the mansion are irreproachable. It would make an excellent habitation. But it is owned by an eccentric old
lady with extravagant notions of its rental value. Besides, no tenants can endure her nagging, and the house
is left vacant. The snows of winter have blown under
the door and through the window cracks. Big patches
of mould have established themselves on the damp
floors. Rats have gnawed holes in the floors and plinths.
An urchin bent on mischief threw a stone through a
window of an upper story, and a heavy spring rain
coming on, the floors are flooded. The plaster cracked
and fell, and the timbers warped and twisted. A seed
fell upon the stone steps and was washed into a crack.
It swelled and grew and the steps were misplaced. The
damage to the building from natural causes amounted
to a couple of hundred dollars in a single year. The
next year it was not quite so bad, but the next year
still, was worse. The house remained vacant and was
saved from becoming a complete ruin only by expensive
repair. Within the fifteen years that it has remained
idle about one-half of the original cost has been spent
for repairs. Is not this building wealth? Is not all
wealth subject to the same law of decay? Is it true,
then, that the capitalist can afford to allow his wealth
to lie idle rather than to lend it without interest? Did
the house grow in value in these fifteen years? If one
had occupied that house during those fifteen years and
merely kept it in repair without paying a cent of rent,
he would have done the owner a very substantial service. The owner would have been spared all outlay for
repairs and would still have a habitation fit for occu-
pancy.

A great mill had been built in a prosperous manufacturing district. The ore which was consumed by the
plant became more difficult to get in that vicinity,
while other fields of supply were opened at a distance.
The ores in the new locality were plentiful and could
be manufactured more cheaply there. The industry
was transferred and the mill first built was shut down.
The doors were closed and the building left to stand.

Twenty-five years passed by. The new mine became exhausted, and the old center of industry revived. The corporation which had shut down its mill years before, concluded to start again. An elder seed had fallen between two heavy pieces of machinery and there taken root in the accumulated soil. As a result, the heavy pieces were displaced and the whole plant thus deranged. In another place, the frosts of winter had caused a wall to cave. The building had become shaky and unsafe for supporting the heavy machinery. In other places, rust had destroyed the fine bearings and weakened the cogs. The plant had, in fact, become a ruin, and but a small portion of the machinery could be used in the construction of a new mill. This was wealth left to itself. Did it grow? Now, if this plant had been kept in operation, as it probably might have been, had no interest been demanded for its use, and had it thus been kept in repair; although the owners had never received a cent for its use, they would have been just the value of the plant better off. They would have been done a very substantial service. This would have been lending without interest. The persons who used the plant would, probably, have been benefited by its use. This would have been reciprocity of services.

We see that the principle upon which interest-taking is founded is not only absolutely false but monstrously absurd. That it is denied by every consideration bearing on the subject; that it leads to injustice and distress. The consequences of interest-taking, as exemplified in our financial history, are ruin and disaster. Capital profits are founded on the same principle as interest, and differ only in the manner in which they are collected. Man is entitled to nothing for which he has not given an equivalent in service; and use of wealth, being paid for by the keeping up of that wealth, is no service for which other remuneration can justly be charged. We then conclude, that neither the landlord nor the capitalist, as such, should receive aught of the product of labor; hence the whole product should go to the toiler.

CHAPTER X.

"That which the hour creates, that can it use alone."

THE argument of Bastiat is considered the argument, *par excellence*, for the justification of interest-taking. In fact, it is the only argument on that side of the ques-tion which goes below the surface. If that argument has not proved interest-taking right, political science has, so far, failed to justify it. The whole gist of Bas-tiat's argument is reciprocity of services founded on the assumption that wealth is essentially and spontaneously productive. The lender, the economist assumes, does the borrower a service by allowing the latter to use his wealth, and the borrower should do a service in return. Paying interest on money borrowed is such a service, and unless the borrower pay interest, Bastiat holds that he returns nothing for the loan.

I contend that the argument is entirely mistaken. Wealth is not essentially and spontaneously produc-tive, as I have abundantly shown. The sheep, without the care of the shepherd, are no more productive of wealth than is the barren daric. It is the care of the shepherd and the food which he provides for his sheep, which makes them productive as wealth. If sheep were as productive when allowed to run at large, uncared for, there would be no object in having shepherds. The fact is that the flock would at once become a prey to its hereditary enemies, the wolves and dogs and the inclement weather, and, instead of increasing, would dwindle away to very small numbers. Then, as soon as they had become wild, they would cease to be wealth, and the labors of the shepherd would be turned into the labors of the chase, and this labor would be that without which the sheep would be unavailable for the satisfaction of human wants. This is true of all so-called productive wealth. Analyze it and we find that it can no more exist without labor than can any other wealth.

Then the payment of interest is not the only service which the borrower does the lender. If the lender does the borrower an incidental service by lending him wealth, by that very act he does himself a far greater service, as it is absolutely necessary to lend wealth to preserve it. The borrower does the lender a very great service by taking his wealth, keeping it for him and returning it to him without deterioration. I have already cited instances.

Bastiat insists, and justly, that money is not wealth, and that we must determine the laws of wealth by considering real wealth, of which money is but the representative. But this entirely disposes of his first illustration of sixpences and crowns. If money is not wealth we cannot deduce from it the laws of wealth. If Paul's sixpence consisted of wealth which he did not wish to use for a year, it would have deteriorated in value at the end of that time. I therefore hold that Peter, or any other borrower, would have been doing Paul a service by restoring to him his wealth unimpaired at the end of twelve months

In his illustration of trading a house for a ship, Bastiat introduces a fallacy which is the groundwork of his plausible but fallacious argument for the justification of interest. The capitalist who actually lends wealth is one who has wealth which he cannot or does not wish to use immediately, or which he wishes to lay by and save for some future time. He is one who has enough for the present, besides that which he lends.

Lent wealth is surplus wealth, so far as the owners are concerned. Bastiat's capitalist is a poverty-stricken laborer who is asked to fold his arms and whistle while another laborer takes his tools and uses them for his own benefit. In lending, these conditions never exist. The wealth which is borrowed, or upon which interest and capital profits are charged, could not have been used in production if it had not been borrowed. Paying one to lie idle or to work with inferior tools, while by the use of his own he might have done better, is something quite irrelevant to the question of interest. Bastiat uses it to cover up the real question at issue. Public policy as exemplified in the common law has long ago condemned the conditions upon which Bastiat's interest argument is founded. Let us consider the real capitalist, the lender of surplus wealth.

Now, if Bastiat should say that a man, after trading a ship for a house, took the house to live in and wanted to borrow the ship for a year; and that the man who traded the house for the ship had another house which he was content to live in, and could not, himself, use the ship for a year, he would have stated the conditions under which loans are really made. The new ship-owner would have been put to no inconvenience in giving up the house, as he would have been left a house good enough to live in. If he did not wish to use the ship for a year, he would wish to have it kept for him that length of time without deterioration in value. If the man who just traded away that ship should assume the responsibility of taking it. and should use it for a year and return it to the owner without deterioration, or in a better condition than it would have been in, had the owner left it idle, I contend that he would thus have been doing the new ship-owner a favor. The owner of the ship would have been relieved not only of the necessity of repairing his vessel, but he would have not even the trouble of taking care of it, and would have it in good order for use at the end of the year. The man who used the ship would also have been benefited by its use. There would have been reciprocity of services, the requirement of Bastiat.

As to Bastiat's third illustration: If Mondor spent his time and surplus cash in building a house to live in and he has no other house, he is not in the position of the lending capitalist. If he have more houses than he can personally use, he gives up nothing in allowing some one else to live in one of them. The house which Mondor cannot or does not wish to use immediately, is surplus wealth which Mondor wishes to save for use at some future time. Such a saving can be attained only by allowing some one else to use the property, and, in return for its use, to keep it in repair. At least that would be the most economical method for Mondor. If there were no borrowers, what would Mondor have done with his surplus house? He might close it up and pay for repairs made necessary by the ravages of mould, rot, rats, etc. That is, in Bastiat's illustration, he would pay the architect $300 per year for keeping his house from becoming a worthless ruin. By giving the house to Valerius for a specified time, he would deprive himself, then, of the opportunity of paying for repairs upon it. If his other house should burn, to be sure, he might have delay in gaining possession of the house in the hands of Valerius, but this is a dim contingency more than compensated by having his house kept in repair. Valerius stands for all borrowers, Mondor for all lenders. It would be entirely irrelevant to say that Mondor might lend to somebody else.

Bastiat thinks that as a first condition of the loan, Valerius should refund the money paid by Mondor to the architect for repair of the ravages of time on Mondor's house. But why should Valerius refund this money? Bastiat says that it is but fair. Why fair? Is Valerius responsible for the ravages of time? Did he make the natural law that houses and all other forms of wealth shall be subject to decay? Do these ravages make the house more useful to Valerius? Why then should he, rather than Mondor, bear the brunt of the law? Bastiat puerilely says that the decay occurs while Valerius is in the house and hence he should make it good. Would it not have occurred to a greater extent had the house been vacant? Finally, when the ravages of time are repaired, who gets the benefit? Mondor, certainly. Mondor, then, should pay the expense of repairs. If Valerius should pay for the repairs as well as pay rent, there would have been no reciprocal service done him for the outlay, and, according to Bastiat's own criterion, Valerius could not be charged with the expense. The advantage of which Mondor deprived himself for the benefit of Valerius is the measure of the service which he did the latter, and the remuneration which he as lender should receive. He deprives himself, at most, only of the opportunity to use his house for a specified time, should a contingency arise making it desirable to do so. Valerius has secure possession for a time, and if for this advantage he refunds to Mondor the three hundred dollars of architect hire, if he stands between Mondor's house and the ravages of time, he more than repays Mondor. Where then comes in the excuse for interest-taking? Interest in this case is commonly called rent, as it is included in the charge for land-rent. Every cent collected for rent is an extortion for which Valerius gets no reciprocal service. If Mondor is paid for what he relinquishes, he has no right to inquire how much Mondor is thereby benefited. If Valerius, while occupying the house of Mondor, had saved it from the flames, there is not a person living who could not see that he had done Mondor a substantial favor, and few would have the hardihood to say that he should pay Mondor for the privilege of so doing because Valerius had by the same act saved a habitation for himself. The case is parallel. It is a beneficent law that he who has most need of wealth is benefited most by its use. It is a relic of the savage in our nature that prompts to want remuneration not only for what we relinquish, but also to appropriate all of the incidental benefit which may accrue to another from our act. It is the old fable of the lion's share. Civilization can give it no sanction.

On the other hand, if rent for the house were collected from Valerius, and Mondor were obliged to bear the deterioration due to the ravages of time, after a few years he would have no house to rent. He could not eat his cake and have it too.

Bastiat's illustration of Malthurin and the sack of corn repeats the same old fallacy. Malthurin, according to the illustration, must have his sack of corn to live on, or he must work for a pittance from day to day in order to keep alive; and in that condition, he is asked to lend his sack of corn to another. What an illustration of a loaning capitalist! If he were a capitalist, he would have had more to live on than he wished personally to use at that time and that sack of corn would represent something which he would have been saving for some future time. It would have been corn additional to his present wants. If Jerome should take this corn, and, at the end of a year, when if stored, it would have been damaged by weevil, damp and rats, should return to Malthurin a fresh sack of corn in its stead, he would have done Malthurin a substantial favor. Jerome would, at the same time, have produced corn for himself. The service would have been reciprocal and Bastiat's requirement would have been fulfilled.

Malthurin would be in the position either of the man who was saving for future contingencies or one who had not yet wealth enough to apply to some purpose for which he was saving it, and wanted that already hoarded preserved for himself until he had procured more. As lending capitalist, he would be like the man who had procured eggs with which to make a cake and wanted them reserved until he had secured the sugar, milk, fruit and other ingredients. If he should store the eggs, they would probably spoil before the other ingredients had been secured. Then, if some one should borrow his eggs and use them on condition that an equal number of fresh eggs should be returned for those borrowed, when the man had secured his other ingredients for his cake, he would have fresh eggs and would have been done an important service.

The illustration of James and the plane is still more fallacious. It jumbles together in James, the rights of capitalist, manufacturer and inventor. The actual loaning capitalist, as such, is an idler with more wealth under his control than he can personally use He

neither invents nor produces. To place James in the
position of the loaning capitalist, we must think of
him as making a plane each year to lay by and sell at
some future time, that he might finally live at ease on
the proceeds, or gratify some other desire. Without borrowers to take his planes, he would have to store them
somewhere to preserve them Rust, rot, worm and mould
would vie with one another in their destruction. When
James wished to sell the planes, he would find many
of them well-nigh worthless. If William should take
the planes and use them and return in their stead good
new planes, would he not be doing James a substantial
service? James would have bright new planes when he
wanted to use or dispose of them, instead of rusty old
ones, as would have been the case had the planes been
stored. Bastiat admits that wearing out within a year
is a necessary concomitant of the usefulness of a plane.
If William pays for that usefulness by supplying a plane
instead of the one which had worn out, why should he
be called upon to pay for it again in interest? There
would be no justice in James having the benefit of the
usefulness of the tool and not be obliged to bear the expense of the wear incident to that usefulness, as well
as the ravages of nature. The fact is, that interest
charges are not founded at all upon the usefulness or
effectiveness of this or that tool. They are founded on
the assumption that wealth, apart from the idea of
the inventor, is inherently productive, which we have
seen to be false.

We see, then, that the loaning capitalist asks the laborer not only to share with him the wealth which the
laborer's toil has produced, but also to make good the
destruction which nature visits on everything produced
by man. This is the essence of interest-taking, yet no
one can give a good reason why the laborer alone should
be held responsible for the inexorable laws of nature.
No one can say why it is just that the laborer should
bear all of the burden for but a pittance of the reward.

The lender sacrifices nothing. The wealth which he
loans is surplus wealth. However potent as an instrument of production in the hands of others, it is useless

in his, for his hands toil not. This fact must be borne
in mind: Unless somebody borrow the wealth of the capitalist, he must stand idly by and see nature steal away
its usefulness. Then the person who borrows that
wealth and saves it from the decay of nature, does the
capitalist an all important service. It is no answer to
say that the laborer at the same time gains an advantage from the wealth which he borrows. Does the gain
of the laborer make the gain of the capitalist any less?
Capital cannot produce, the laborer can. The laborer
has lived without capital, without wealth except the
strength of his muscles and the cunning of his brain.

With this strength and cunning alone to start with,
the laborer has wrested from nature all that there is
of wealth in the world to-day. Destroy every vestige
of what men call capital and enough people would survive the calamity to repopulate the world and reorgan-
ize society. Destroy the power to work and in ten
years there would not be a living human being.

Nobody who considers what man has sprung from
will deny this. Man did not come into a world of
walled cities, palaces and machines. He was once a
shivering naked savage, his implements clubs and stones.
His bread he plucked from the trees by labor, his meat
by labor he pursued and killed. Man always earned
his bread in the sweat of his own brow or the brow of
somebody else. The laborer has the producing power
of nature; capital, the decaying principle of wealth.
Why then can it not be confidently asserted that the
laborer, apart from nature, exerts the only productive
force? The laborer can put his stamp on the treasures
of nature's storehouse and the product is wealth. Nature will not receive the stamp of capital.

Thore asks: "Will an extra crown appear at the end
of a year in a bag of one hundred shillings? Will there
be two hundred shillings in the bag at the end of fourteen years?"* No, nor an extra grain in a bag of corn
(Bastiat to the contrary notwithstanding.) Herds will
not increase without the laborer's care; fields, untilled,
will not yield a harvest. All of Nature's favors must

, *It will be remembered that Bastiat said crowns would not grow, but corn
would. Yes, with labor.

be wrested from her by the arm of toil. Where then is
the justification of interest? From whatever point we
view it, we can see no warrant for the assumption that
wealth has the power of spontaneous growth. As we
have seen, if wealth does not grow spontaneously, there
is no ethical basis for interest. Each man has a right
to what he produces and no one has a claim on what
is produced by another.

Let us allow the great apostle of interest himself to
tell of the advantage which the capitalist has over the
laborer. We have examined his reasons for believing
that that advantage is just. Here is a translation of
Bastiat's own words:

"Here are two men, one of whom works from morn-
ing until night from one year's end to another, and if
he consumes all that which he has gained, even by superior energy, he remains poor. When Christmas comes
he has no more ahead than he had at the beginning of
the year, and has no other prospect than to begin again.
The other man does nothing either with his hands or
with his head; or, at least, if he makes use of them at
all it is only for his own pleasure. It is allowable for
him to do nothing, for he has an income. He does not
work, yet he lives well, having everything in abundance, delicate dishes, sumptuous furniture, elegant
equipages; nay, he consumes daily things which the
workers have been obliged to produce by the sweat of
their brows, for these things do not make themselves,
and as far as he is concerned, he has no hand in their
production. It is the workingmen who have caused the
corn to grow, polished the furniture, woven the carpets.
It is our wives and daughters who have embroidered
these stuffs. We work for him and for ourselves. For
him first, and for ourselves if there is anything left.
But here is something more striking still. If the former
of these two men consumes within a year any profit
which may have been left to him within that year, he
is always at the point from which he started, and his
destiny condemns him to move incessantly in a perpetual circle and monotony of existence. But if the other,
the 'gentleman,' consumes his income within a year, he

has the year after, in those years that follow, and throughout all eternity an income inexhaustible, perpetual. Capital then is remembered, not only once or twice, but an indefinite number of times. So that at the end of a hundred years a family which has placed twenty thousand franks at five per cent interest will have had one hundred thousand franks, and this will not prevent it from having one hundred thousand more in the next century. In other words, for the twenty thousand franks which represent its labor (or the labor of some one else), it will have a ten-fold value in the labor of others. In this social arrangement is there not a monstrous evil to be reformed?

"And this is not all. If it should please the family to curtail their enjoyment a little—to spend, for example, nine hundred franks instead of a thousand, it may, without any labor, without any other trouble than that of investing the other hundred franks a year, increase its capital and its income in such progression that it will soon be able to consume as much as one hundred families of producing workers. Does not this go to prove that society is nursing in its bosom a hideous cancer which ought to be removed at the risk of some temporary suffering?"

Yes, Bastiat! it certainly does, and your illustrations of planes and ships and sacks of corn, although they may obscure the seat of the terrible disease, cannot hide its manifestations. The skilled social physician can see through your thin mystifications. He would be obtuse, indeed, who could not see a wrong in rewarding one person a hundred times more than another for a given service. He would be still more obtuse who could not see the fallacy in the assumption that one who hoards a given amount of wealth and lends it is entitled to remuneration for the act *ad infinitum*.

You assert that the twenty thousand franks represent the labor which that family has performed. This may or may not be true. Many of our modern fortunes represent wealth appropriated from the toil of others. They may represent labor, or may be the result of labor, but as we have seen above, they are not labor and have

none of labor's productive power. Labor's result is
wealth, nothing more, useful for consumption, or to
the laborer in production, but of itself subject only to
decay and destruction. Why it should claim remuneration is not clear, unless for its gradual but no less
effective disappearance from the earth. And, after all,
to speak of remunerating either capital or labor for
services is utterly absurd. Inanimate things have no
claims on man.

But granting for the sake of argument that these
twenty thousand franks should earn for their owner a
remuneration, on what ground of right or justice should
they earn a greater remuneration than twenty thousand
franks' worth of the labor of the toiling citizen? Why
is the capitalist's fortune more worthy of return than
the twenty thousand franks which have gone to the support of the laborer and his family. The worker's
strength must be kept up by constant feeding. The
wealth of the capitalist is almost as perishable, it must
also be kept up by constant accretion. The labor produces more wealth from the strength which he obtains
from the food which he consumes. The owner of the
franks produces no wealth, neither do the franks. Let
them lie idle in a vault and they would not increase
one jot for all eternity. Store the real wealth, represented by them, and you would have nothing left at the
end of a score of years. Why then should we remunerate the owner of the franks not for one year alone but
for all eternity, while the wealth which the laborer receives is gone when consumed and we remunerate him
but once. We go further and not only compel the
toiler to make good to the owner of the franks the ravages of nature, but also to pay him a hundred-fold for
what he produced, inherited, or perhaps obtained by fraud
or force.

Why should we place such a premium on the saving
of wealth and reward its production so little? Is accumulation the end of civilization? Here again we run
counter to natural law. Natural law says that hoarding wealth beyond our needs is criminal folly, and im-
poses the heaviest of fines on miserly instincts. All the

progress made by man has been in the direction of hoarding less as compared with current consumption. All improvements in machinery, all improvements in methods of industry, have been in the direction of giving man more to use and less to hoard. And it is necessarily so. The more we hoard the larger a percentage of our annual product must be devoted to the keeping up of the wealth already hoarded, and the less we will have to use. Under such conditions undue hoarding would lead us to such a point that nearly all our energies would be spent in keeping up the wealth already gained, and while our store would be large, we should be unable to satisfy our current wants.

Nature rebukes hoarding in most unmistakable terms. Her fines are levied on every dollar's worth of wealth that is stored for future use. She declares that the whole effort of man, if he would progress, must be in making his product agree in comparison with his capital, and in reducing his surplus to what is absolutely necessary in supplying capital for the increasing population. The yearly penalty in our present hoard is about three billion of dollars, or one-fifth of the wealth produced. It behooves us to make that hoard as small as possible, while retaining the effectiveness of our industries.

Yet in the face of this, we say to the world: "You who have saved even so much as a laborer produces every five years of his active life, can live all the rest of your days in idleness, if you so desire, and your children and your children's children may do the same. He who has not been fortunate enough to save must divide with you his substance even to keeping you in idleness. He must toil unceasingly, and when he shall have been gathered to his fathers his children after him must toil, and a portion of everything which they produce belongs to you and yours by the right which your saving gave you. And the toilers must not be niggardly about feeding you. Your share shall every fourteen years equal your original saving, and yet your fortune shall never grow less. By the simple act of saving a fortune, insignificant as compared with what one laborer produces during his lifetime, you have removed from yourself and posterity the curse of humanity! Man must eat his bread in the sweat of his brow? Or, perhaps, your father has done it for you; perhaps an uncle, perhaps a more distant relative whom you never saw has left you a small amount of perishable wealth, and by that act saved you from the necessity of laboring, made you a sharer in the results of other's toil. By a sort of vicarious industrial merit you have been made a favored one of earth.

To the man who produces unceasingly, but who cannot or does not save, we say: "You must stay nature's destroying hand. The substance of the capitalist is sacred; see to it that you preserve it. Keep it replenished after the waste of time, and besides see that he has enough to live on. Then, if there is anything left, you may take it as your own. Hoarding, you must remember, exempts forever from toil; mere producing gives only the right of sharing that production with those who toil not."

These are the speeches which we act out when we sanction the practice of interest-taking. No questions are asked as to how the wealth was hoarded. Its possessor is virtually pensioned for all time and billeted on the community. Interest rewards the capitalist *ad infinitum*. It is wrong. If for the twenty thousand franks representing the laborer's toil he is remunerated but once, the twenty thousand franks which represent the capitalist's earnings should gain for the capitalist but one remuneration. That remuneration would be the right to consume wealth to the amount of twenty thousand franks, nothing more. In the light of logic and ethics no other conclusion is possible. Men have equal rights.

It will be seen how absurd this idea of interest is, if we consider that it never could be generally applied. If all were to save and lend their money out, or place it where it would produce, all should be able to collect interest, and to live without engaging in actual production. If one can live idle by the productive power of wealth, all who save wealth should be able to live idle.

But we readily see that if all saved wealth, there could be no interest collected and the wealth saved would be useful for consumption only. All would be obliged to work to live. We cannot conceive of a paradise of universal idleness. The wealth would be found to be unproductive. If it would be unproductive then, why not now? It is a false principle which cannot be generally applied. Try to apply it generally and the result will show the unproductiveness of wealth. Wealth is unproductive now, but it may be used as an instrument to take what others produce. This is what interest-taking means.

Bastiat has well said that things do not make themselves, and that the capitalist has certainly no hand in their making. He might have added that neither did the wealth which the capitalist saved produce these things. Leave it unattended and it could not have preserved itself from destruction. The capitalist, then, had no right to take these things from another. The wealth which he had produced had disappeared years before under the inexorable law of nature, yet he is still living on it. What an anomaly! Can one eat his cake and have it too? The capitalist does, but he is the only example of the occult phenomenon. Then it is but a trick. He steals the cake by legal jugglery from the mouths of its rightful owners, and by pretty fictions convinces them that it is his own. Better than the lamp of Aladdin, better than the magician's wand, even better, far better than the philosopher's stone, is the economic fable by whose potent alchemy the possessor of a little hoarded wealth can multiply his gold *ad infinitum*, and levy contributions on the generations of men to the end of time. And it is a magic capable of transmission without the trouble or pains of study. Its adepts are legion. By its action their posterity are made pensioners on all the generations of men. Under its fecundating influence the capitalist's wealth becomes the fabled cup, that, however often drained, is forever full; it becomes the purse which always contains a dollar. Verily the secret of the capitalist is better than the power of kings.

But, like all necromancy, when unveiled, interest-taking is but the jugglery of the faking charlatan. When the wealth which he has saved is gone, the interest-taker mystifies others and takes their wealth to supply its place. It is by the toil of others that the cup is kept full. He shuffles the empty vessel into the place of the brimming goblet, which he drains in turn. His magician's wand is but the barbarous custom of tribute which changes not but directs the stream of wealth from the hand of the toiling producer, into the coffers of the idle parasite. It has obtained so long that men have forgotten to resist it. Interest, the all-powerful necromancer, is founded on the monstrous assumption that wealth has within it the power of spontaneous growth. There is no mistaking the conclusion that interest-taking is wrong.

CHAPTER XI.

A sane man will not wrong himself because he is not allowed to wrong another.

IT is asserted that without the practice of interest-taking, there would be no saving, that all capital would be destroyed, that we would be hampered in our production and retrograde towards the savage. Does our civilization depend for its existence upon the thoroughly barbarous principle of tribute-taking? You may as well say that without gluttony there would be no eating; because one is not allowed to gorge himself, he will refuse to take nourishment to sustain life. Why would there be no object in saving if we could not collect interest? If I produce more wealth than I can use at present, and want to save it for use at some future time, will it not be as much mine when I want to use it, if I lend it without interest, as if I collect ten per cent interest for its use? The contracts for the return of capital and the paying of interest are in no way interdependent. One can be made without the other. I can, as now, make an agreement with the borrower, that if I allow him to use a portion of the wealth controlled by me, he will return it to me at the end of a certain stated period in as good a condition as when I lent it. That is, he will make good the ravages of time. It is not necessary for me at the same time to contract for interest, and if the agreement is carried out, I will be sure of getting back all that I have produced. This is as great an incentive to saving as any mortal would

75

require. One would, as now, look forward to a time of ease, when he might live on what he had saved during his life of active production. He would be obliged to lend his wealth in order to save it. The same security would be required as under the system of interest-taking. In fact, loans would be much more secure, for toilers would have reduced burdens to meet and be more prosperous, and it is a well-established principle in business that collections are easy in times of prosperity. The argument of no motive for saving unless interest is allowed, implies that humanity is so avaricious that if one cannot get what does not belong to him, he will not care for what he has. Self-interest will see to it, that enough is saved to keep things moving. Men usually, unless checked, take all that they can get, but are satisfied with what they can get, unless they can get more. Under no interest, each would get what he deserved, no more.

Under the system of no interest, men will not get rich while they eat and sleep and loaf or debauch, as at present. Statisticians cannot fill pages with calculations showing how much richer each succeeding breath finds a Vanderbilt or an Astor. Mere existence will not be a wealth-collecting enterprise. As soon as one ceases to engage in productive toil, his fortune will begin to grow less and decrease each day by just the amount which he spends. He will have all that he produces to use as he pleases, but he must take his hands off the production of others. To say that under such conditions men will not save, to say that men must have more than they are entitled to before they will take care of what they have, is like saying that rulers will not govern unless the people give up to them all popular rights. As much as royalty has been curtailed, they are still not anxious to give up what is left.

It is quite true that under a system of no interest there would be less lending. There would be fewer borrowers, for every one would be given his own. Then men would personally use more of their own wealth in employing themselves. But instead of being undesirable, this would be a great advantage. There would be a much larger number of productive workers, much more wealth produced, and hence a larger share for each consumer. Each could more readily employ himself or take care of himself than now. There would be less motive or necessity for borrowing. But there would be the strongest motive for "laying by" something for declining years, a motive strengthened by the knowledge that what one produces will not be taken from him by idlers. Each man has the strongest motive, too, for keeping his surplus wealth in the hands of some one who will preserve it for him. For, as we have seen, wealth cannot be hoarded with impunity.

Another will arise and say: "Suppose the capitalist refuses to lend his wealth. Suppose the owners of houses, for instance, refuse to let them be used, how will the poor be sheltered?" The case is impossible. Suppose all men were insane, we should have a terrible world; but they are not, and to argue from such a condition is starting with a false premise. Experience teaches us that men will, on the whole, do that which they believe to be most conducive to their interest. If they can get something for nothing they will take it; if not, they get it as cheaply as they can. If a man had two thousand dollars invested in a house which he could not use himself and for which he had no prospect of collecting rent, rather than allow that house to go to ruin and the two thousand dollars to be lost he would either let that house on the condition that it be kept in repair or he would sell it. If he wished to sell it, he would be obliged to sell it to those who wanted it, and on such conditions that they could buy it. If no interest were charged on the purchase price, any renter could, within a few years, buy the house which he occupies by merely applying to the purchase price, rent and interest charges now paid. In that way the bulk of houses now built would continue to be occupied as now, although many might change ownership. As for the habitations for the increase of population, they would be similarly provided. There would probably be surplus wealth and it must be held in some form. There is no better form for the preservation of wealth, bullion excepted, than that of buildings.

To the producer, the discontinuance of interest-taking would be an unmixed blessing. He would be enabled to use the wealth which its owners could not use, and with it produce more wealth. He could, at the same time, make up for them the inevitable destruction visited on wealth by nature, and increase his own substance. He would be released from the hard conditions which now, so often, make production unprofitable to all except the money-lender. The burden on business which now sends the country into practical bankruptcy every decade, and makes ninety-five per cent of all business men fail, would be removed. The toiler would not be obliged to hand over his substance in interest to these who toil not, and would be able to accumulate a surplus of his own or to shorten the hours of toil. There would be no drones among those, capable of working. As soon as one refused to work, his fortune would begin to melt away, and even if the amount of the wealth he had accumulated were up into the millions, instead of multiplying, as at present, it would begin to slip from the clutches of the idler It would be only a question of time until the fortune, however large, would be exhausted, and the idler and his descendants would be obliged again to take up their burden with the rest of mankind. The accumulated fortunes of the more fortunate would be amply sufficient to supply their declining years, and there would be enough also to educate their children and to give them a start in life; but they would not grow richer than their fathers unless they worked and added something to the wealth of the world. The worthless, idle scion of a wealthy family, would be a thing unknown. No fortune would be sufficient to bear his extravagances for a lifetime. He could not indulge every luxurious whim. Once amenable to the benign, unshackled law of nature, that man must eat his bread in the sweat of his brow; he who inherited a fortune would grow poorer and poorer, unless he produced, until finally he would be obliged to work, beg or starve. All idlers, rich and poor, would be placed on an equal footing. The wealthy idler could not save himself by refusing to lend. If he tried to hoard his

wealth, nature would punish him by destroying it all
the more rapidly.

And this, as we have seen, is the only way in which
the toiler can be benefited. There is not enough produced for him, the interest-taker and the landlord. He
is entitled to the entire production after keeping up
the fixed capital of the country. If he would increase
his wages, he must insist upon his rights, and do away
with interest-taking, and appropriate rent. Until he
does this he will have to bear a constantly increasing
burden, which will finally crush him to the earth. If
burdens are to be made lighter for toiling shoulders,
more shoulders must support these burdens.

CHAPTER XII.

"By their fruits ye shall know them."

INTEREST-TAKING, then, is wrong. It has no warrant in ethics. It contradicts natural law. Its logical consequences are absurdities. Its basis is an untruth, its practice an imposition. One single dollar collected that way is a dollar extorted, taken without a shadow of return. But the enormity of the amount extorted makes interest-taking a most potent factor in economic disorders.

The scientific test of an inductive theory is the number of observed phenomena which can be explained by it. There are a number of the observed phenomena of economic science still awaiting explanation.

1. The cause of industrial depressions and financial panics manifesting themselves periodically and extending to nations with all sorts of government and revenue systems, but all of whose financial systems are founded on rent and interest-taking.

2. The extremely rapid accumulation of wealth in the hands of a comparatively few non-producers.

3. The abject poverty of a large percentage of the producing masses.

4. The failure of improved machinery to better the conditions of the producing masses in a degree at all commensurate with the increased producing power which it has given to the laborer.

5. The fact that non-producers receive much the largest salaries.

6. The fact of so many laborers being doomed to

involuntary idleness at times when they are most in need of employment and there is most need by others of the goods which they might produce.

7.   The fact that work is looked upon as a boon, and in times of greatest industrial depression the most extravagant undertakings seem to promote prosperity.

8.   The fact that destructive wars often stimulate industry and open up to nations new eras of prosperity, although both life and treasure have been lavishly squandered, and the laboring force of the nation made less.

9.   The tendency of unwarlike nations to fall into industrial and political slavery.

10.   The tendency of lavish wealth to undermine the integrity of the nation and lead to its decay and final destruction.

11.   The decay of the American yeoman farmer, the unprofitableness of his toil and the consequent tendency of rural populations to drift to the cities.

12.   The heavy and rapidly increasing mortgage indebtedness of the country.

CHAPTER XIII.

False principles lead to evil results.

It is self-evident that financial and industrial panics show derangement in financial or industrial systems or both; just as certainly as sickness shows derangement in the animal system. This periodical derangement in financial and industrial systems has never before received a satisfactory explanation, if indeed it has ever received any explanation at all. The taking of unearned incomes made possible by the private collection of rents, and the collection of interest and capital profits fully explains these panics.

The explanation is that they who produce are not allowed to retain enough of their product to pay necessary current expenses, and they fail. When many fail at the same time there is a panic.

Interest is not a necessary expense of production. Neither is rent; at least from the standpoint of the whole people. If capital were lent without interest it could be used to just as much advantage; if the rent of land went to the whole people, land would be just as productive.

It is not difficult to show that the rent and interest charges on the country's industry are too large to be met, and at the same time pay the current expenses of production. Under these conditions business men must fail and allow their property to pass into the hands of others. General failure produces panic.

Interest and rent charges, added to royalties and spec-

ulative profits, amount to three and one half billions
or more per year.

We have available to meet this charge but about a
billion and one-half per year. This is cause enough for
financial panic. A few estimates will show that this
is not a hollow assumption. According to Poor's manual,
the rents and capital profits paid on railway property
in the year 1891-2 was, in round numbers, four hundred
and eighteen millions of dollars. According to the ab-
stract of the Eleventh Census, mortgages paid in round
numbers in 1889 four hundred million of dollars in interest. According to the same authority, the manufacturing capital of the country amounted in 1890 to six
and one-half billions, and it paid an interest charge of at
least three hundred and ninety millions. That capital
pays an interest and capital profit of more than eight
hundred and fifty millions, so that after allowing liber-
ally for the amount of manufacturing capital included in
the item of mortgages and deducting enough to meet all
future duplications in figures, we would still have three
hundred and ninety millions additional paid upon it in
interest and capital profits. The mining capital of the
country is estimated to pay a capital profit of two hundred and six millions per year. At six per cent it would
pay a charge of over seventy millions annually. The
public debt pays an interest charge of ninety-five and
one half millions. The capital employed in retail and
wholesale business pays a capital profit or interest of
not less than three hundred and sixty millions per annum. The house and office rent of persons who live in
hired houses, exclusive of the rent of land on which
these houses are located, is not less than two hundred
millions per year. Bank discounts in commercial transactions are far up into the millions, but cannot here be
estimated with any certainty. Clearings of sixty billions per year would indicate discounts of hundreds of
millions.

The banks show a capital profit, net, of sixty-seven
millions per year. Hired farms pay on improvements,
which is, strictly speaking, an interest charge, fully
twenty-five millions per annum. Insurance, fire and

life, collects an interest charge on its capital, of sixty-
five millions. Fisheries pay an interest charge of two
and one-half millions.

Lumbering capital pays an annual interest charge of
not less than thirty-five millions. Then there are large
miscellaneous interest charges that I shall not attempt
to estimate. The aggregate of these already named is
not less than two billion thirty-three million. Add
to this three quarters of a billion, the amount charged
annually for rents and not included in the above,
and a quarter of a billion or more for royalties on
mines, oil wells and timber, then add speculative
profits, and we will have at least three and one-half
billions unearned charges to pay yearly. Considering
the fact that bank discounts are not included, the estimate is extremely modest. These, as we see, are largely
composed of interest proper and capital profits.

This may be arrived at in another way. The wealth
of the United States is estimated at something more
than sixty-five billions in 1890. Homes owned by their
occupants, a portion of the farm property of the United
States, and the property of some very small retailers is
all that escapes paying interest and capital profits. Fifteen billions would be a large estimate for this non-
interest and rent paying wealth, leaving about fifty
billions to pay interest and rent. It is a well known
rule of economics that where any capital used in business tends to collect interest, all capital used in business will tend to so collect interest, for if the business
man could make as much by loaning his capital as he
could by employing it in business, he would not take
the risks of business. On that principle we see that our
estimates of interest and rent charges are not too large.

Besides the interest, rent and speculative charges
above estimated, we pay the creditor class an average
of one billion or more per year, on account of the appreciation of our money standard. Within the last
twenty years our money standard has appreciated, as
compared with staple commodities, between thirty-five
and forty per cent. This is denied by no intelligent
person, and would involve the yearly payment to the

creditor class of fully the sum named above. But we need take no notice of this in our present estimates, as we are calculating on a gold basis.

It is conceded by all authorities, that after paying the current expenses of production—wages, repair, necessary improvements in machinery, deterioration, etc., ten per cent of the gross product is a very large estimate of the margin of profits, including rent and interest. The gross product of the industries of the United States is not more than fifteen billions yearly, and hence the amount available to pay rent and interest is not more than a billion and one-half yearly. This will fall short by two billions of meeting the demands upon it.

Even without actual itemizing it may readily be seen that the interest charges of the country's industries are enormous. An attempt is made to collect interest on every dollar invested in business and on every dollar representing debts unpaid. And this must be paid out of the gross product of each year. From each year's product, the wealth of the world must be kept up. Buildings, machinery—everything must be kept in repair. Improvements for use in the future must be taken from the stock of the present. The increasing population must be supplied. There is not enough wealth produced to meet all of these obligations. Either the current expenses of production cannot be paid or the fixed charges of rent and interest cannot be met. If current expenses are not paid, manufacturing plants deteriorate, fixed capital is encroached upon, wages are reduced and laborers thrown out of employment. Current obligations are not met. The business man finally becomes a bankrupt, or the wage-workers become bankrupts and outcasts depending on charity for support. If interest is not paid, then the wealth hypothecated for the loan is appropriated by the lender, and the borrower, failing to meet his obligations, becomes a bankrupt. If rents are defaulted the natural resources of the country are withdrawn from use in production. He who had been using them becomes a bankrupt and industry is prostrated. When all expenses of production exceed the gross product, the industries of the country are not paying as a whole. If that condition of things continues, the industries of the country or a large percentage of them will bankrupt. A majority of the business of the country must go into the hands of a receiver, pay up a percentage of its debts and make a new start.

The country with all of its allied industries is analogous to a mammoth business concern. When it contracts greater liabilities than it can meet, it fails, and we have a financial panic. But unlike a mammoth business concern, the industries of the country do not bankrupt as a whole. There is always a considerable percentage of the individual industries of the country which pay a profit even in time of panic. These are gaining ventures while the average of all business will show a balance on the side of loss. These survive panic and others fail. A failure of the industries of the country to pay as a whole is made manifest by the failure of a large percentage of the industrial institutions of the country. This state of bankruptcy is chronic. Counting everything, the aggregate liabilities of the individual industrial establishments of the country are always greater than their assets. The majority of individuals of the industrial world are always in a state of potential bankruptcy, and it is credit alone which stays the granting of the receivership. Men in business become involved as debtors and creditors. Mutual confidence in the meeting of all liabilities tends to keep business moving for a time, and industries seem prosperous. Any disturbing of credit or shaking of confidence precipitates a financial crisis. There is a panic, numerous failures, liquidations at a discount, and business goes on anew until the obligations become obviously larger than the wherewithal to meet them, when another panic ensues. The failures recurring every day act as a sort of safety valve to put off the time of panic.

These are the conditions which lead to financial panic. It is nonsense to say that any considerable portion of the business concerns of the country would regularly and necessarily fail unless their liabilities were greater than their available assets. This is the only condition under which a failure can occur in one case or in ten thousand; in a country grocery or a manufacturing establishment with an output of millions. What brings about this condition?

It is the foundation stone on which our industrial system is built. The basic principle of our industrial system leads us to recognize obligations which we can never meet. It is the principle which asserts that a dollar will grow into two dollars in a number of years and keep on multiplying until it represents all of the wealth on earth. It is because we try to pay a rent and interest charge of three and one-half billions with a net product or margin of profits of about a billion and one-half. It is because it costs us, besides rent and interest, about thirteen and one-half billions of dollars to yield a product of fifteen billions of dollars, and we try to charge that product with three and one-half billions of dollars in unearned charges also.

There is no fiction in these statements. According to Atkinson, a successful business man makes a profit of but six per cent on his investment, including interest. According to the same authority, a successful manufacturer receives less than five and one-half per cent of the gross product of his establishment. According to the Census Report for 1890, the margin of profits, including interest, for successful manufacturing establishments was nine and seven-tenths per cent. This,

My estimate then has given the benefit of all doubt to the other side, and yet it shows that we fall behind about two billions per year in meeting obligations, under a rent and interest system.

Now let us take into consideration that more than ninety per cent of those who engage in business fail; that they make nothing whatever, but rather lose. When we consider that a vast number of those who aid in production are so underpaid that they actually want for the necessaries of life, and that nearly a mil-

lion of persons in this country are on an average involuntarily idle and unsupported by industry, we will
conclude that if losses of the unsuccessful ventures were
balanced against the profits of the successful ones, the
margin of profits above necessary expenses would be
reduced very materially. But even place that margin
where the census figures for manufactures place it, at
about fourteen per cent on the capital invested or ten
per cent of the gross product, and we would have but
about a million and one-half left after the current necessary expenses of production are paid, exclusive of rent,
interest, royalties, capital profits, etc. Now the
charges for these things are about three and one-half
billions, leaving the small annual deficit of two billions, which is not met and hoards itself up for the future financial panic. This estimate is made on the
basis of a gross yearly product of fifteen billions for all
of the industries of the United States, and this I believe to be a close estimate.* It has been placed, to be
sure at twenty billions, but no figures which I can
obtain will bear out the latter estimate. I can account
for it only on the supposition that many products are
counted twice, in their form as raw material and in the
finished product. The finished product alone should
be counted in estimating the gross product of a country's industries, except such product as is exported, for
wealth is useful for consumption only in its final form.

Thus we cannot add the cost of the wool in the coat
to the final cost of the coat, for the wool has already
been included. In the same way the value of wheat is
included in the flour (except the wheat exported), the
value of the ore is included in the finished machinery,
etc. Then, when we consider that manufactures amount

---

*According to the Census figures, the gross product of manufacturing establish-
ments for 1890 was 9,370 million dollars; the gross products of agriculture, 2,460
million dollars; the gross product of mines, 587 million dollars. From this must
be subtracted 5,159 million dollars, the cost of raw material used in manufactures.
This would leave but 7,258 million dollars in tangible wealth, produced in the
United States in 1890, the only important item left out being lumber. During
the same period the cost of transportation was 1,205 million dollars. This was
of course wealth produced, or at least two-thirds of it, the actual cost of operating.
Then personal and professional services and mercantile services in getting the
goods to the consumer could not possibly have amounted to more than six and
one-half billions within a year. That would show that the middleman got nearly
as much as all others combined, and yet would place fifteen billions as the limit
of yearly production of wealth in the United States.

to less than ten billions annually, including all of this raw material, and farm products to less than three billions, including that sold to manufacturers as raw material; and that the product of mines is worth half a billion per year, nearly all of which is included in the cost of manufactures, it is difficult to figure out a gross product of more than fifteen billions. We have seen that not more than a billion and one-half of this can be paid in unearned incomes without trenching on the necessary expenses of production. But there are fixed charges to be met of more than double this amount. For the last decade we have been paying fixed charges, as the railways put it, on an average of three and one-quarter billions per year. This would aggregate thirty-two and one-half billions in a decade, with but fifteen billions that could be applied with impunity to its payment. This is all that could be spared after paying current expenses of production other than rent and interest. There is an enormous deficit. The consequence is that we do not pay current expenses of production. Business men fail instead, and they alone are paid whose claims are secured. The money-lender, the landlord and the royalty collector, all have their claims secured, and the laborer and the active business man are they who lose and suffer. If an accounting has been put off for a decade, the balance against the active business man has become so great at the end of that time that he is no longer trusted; he is pressed and fails. This is why business panics occur, and the sole reason why business panics occur., We try to pay unearned incomes to a greater amount than we are capable of.

Another consideration will show that industry is unable to bear the burden imposed upon it. According to the figures of the census for 1890, the wealth of the United States has within the last decade increased twenty-two billions. This estimate is made in British gold and ignores the appreciation of the money standard. This would be a yearly increase of about two and one-fifth billions, an increase of nearly three-fourths of a billion over the figures deduced by estimating from

the observed margin of profits.* Now the charge against
this would be an average of three and one-quarter billions per year, leaving a very large deficit.

But it may be said that there is no necessary connection between the annual saving of the industries of the
country and the amount which they are able to pay in
fixed charges, on a solid financial basis. The fixed
charges, like the current expenses of production, are paid
out of the gross product of fifteen billions. That is
quite true. There need not be any connection between
the savings of the country and the ability to pay fixed
charges, but as a matter of fact there is. The simple
reason is that at least all that is consumed is really necessary for the payment of current expenses of production
other than fixed charges, or the charges of rent and interest. If that is true, only the wealth saved can be
applied to the payment of rent and interest without
trenching on the necessary current expenses of production, leading to repudiated obligations, distrust, panic
and bankruptcy. It is not difficult to show that all
that is now saved is all that could be saved after paying current necessary expenses of production other than
rent and interest. As a general proposition, what a
people actually spends is the best measure of what it is
necessary for that people to spend. There is great and
expensive luxury in the country. That might be dispensed with. But will any one assert that what is now
spent in luxury might not, if indeed it should not, be
spent in supplying the wants of toilers? We have a
million devoted to chronic idleness. Would it not take
a goodly sum to pay such wages for such hours of toil
that with our present product, that million would find
a place in the industrial world, and become self-supporting, self-respecting citizens? We have other millions living on short rations. It would take another
goodly sum to pay them wages for their toil that would
give them a competency and make them self-respecting
citizens of a self-supporting republic. There are obligations amounting to millions unmet, repudiated every

*This discrepancy may be accounted for by the fact that the estimate of the
margin of profits did not take into consideration the savings of wage-earners.

year. Those who hold them, often render a service for which they get no return. Would it not make a hole in the luxury account to meet all of these honest obligations? Everything points to the conclusion that what is saved annually in the United States and appears as an annual increment to the national wealth, is the maximum amount that can be spared after paying the current necessary expenses of production, other than rent and interest, and is hence the greatest amount which can be applied to the payment of rent and interest and other unearned charges without leading to bankruptcy, widespread in proportion to the amount by which these fixed charges exceed the net produce of the nation, or the yearly increase of national wealth.

I mean by the necessary charges of production, the charges without which production could not be continued in its present effectiveness. The fact of a landlord collecting rent on a certain piece of land, makes that land no more productive. The fact of a moneylender collecting interest on the wealth which he lends, makes that wealth no more useful to the laborer. If the land and the wealth could both be secured without rent or interest, the laborer could with them produce as much new wealth as though both interest and rent were being collected. But the charge for labor is different. If the laborer does not receive his wages, he cannot get food to keep up his strength and he becomes unable to produce. If he receives inadequate wages his powers of production are impaired, his children are starved and grow up weak and ignorant, poor industrial workers, poor citizens, expensive as invalids or criminals. Productiveness cannot be kept up to its original standard of effectiveness. Then in manufacturing, machinery must be kept in repair or the plant will lose in effectiveness and finally become worthless. Raw material must be paid for or production cannot continue. In farming, land must be fertilized, seeds procured, machinery kept up, etc. In both, taxes must be paid to secure the protection of government. These are expenses which must be paid, and these are the expenses which calculations both from the

margin of profits and the annual savings of wealth
show to fall but about a billion and one-half short of
the gross product of the country's industries for each
year.

I say that two billions a year is above the maximum
amount which can be applied to interest and rent paying in this country without leaving necessary expenses
of production unpaid, and I say it advisedly. While,
according to the census, the country makes a net saving of two and one-fifth billions per year, nearly half a
billion of that is saved by wage-earners and certainly
cannot be applied to either current necessary expenses
of production or interest and rent charges without being re-borrowed. I use wage-earner in the broad sense.
Those persons, professional or otherwise, who are not
rich but receive comfortable salaries for real services
rendered. I base my calculation on deductions from
the census figures of distribution of wealth. If ninety-
one per cent of the population, including wage-earners,
own but twenty-nine per cent of the wealth of the
country, they cannot in the past possibly have saved
more than twenty-nine per cent of the wealth saved,
while the other nine per cent or wealthy portion of the
population must have saved at least seventy-one per
cent of the annual saving. As the wage-earners had
comparatively a much larger share of the wealth of the
country in 1880 than in 1890, between 1880 and 1890
they must have saved less than the above named proportion, which would leave their saving but about half a
billion in two billions, added to national wealth each
year

Thus if A had one thousand dollars and B eight hundred dollars in 1880 and they both together saved one
thousand from 1880 to 1890, and in 1890 A had eighteen
hundred and B one thousand dollars, it would show
that B saved but one-fifth of the total saving for the
decade, although his property was more than one-third
of the whole.

Then only what is left after deducting the savings of
wage-earners from the total national savings, can be
applied to the payment of interest and rent without

trenching on necessary expenses of production, and leading to financial embarrassment. Thus we arrive at substantially the same conclusion as that reached by estimates from the margin of profits.

Fortunes go on piling up under the law of interest, and after all checks and counter tendencies have been negatived, we have a trade depression every ten years or oftener and a panic every twenty years. The fact is, a financial flurry can be produced any time the creditor class demands money, for there are not available assets to meet their demands and at the same time keep business moving. Money, the only payment accepted by the interest-taker, is nearly always massed in the hands of the creditor class or can be collected there on the slightest provocation. In times of confidence, business is kept moving by shifting liabilities, but in times of doubt and uncertainty, from whatever cause brought about, much of the business of the country finds it impossible to meet its obligations and files into bankruptcy. The cleverest speculator cannot long keep his business moving by borrowing from one to pay another, unless debts are very small as compared with the business done. Just so with the majority of business men. The piling up of debts always ends in collapse, and their interests are so interlinked that the fall of one brings down a hundred. It is nonsense to say that lack of confidence is the cause of financial panic. Unless the ground principles of business produce instability, want of confidence can have no effect. Men realize that a great number of the business undertakings of the country are not able to pay what they have undertaken, and they therefore lose cofidence. A building never fell through lack of confidence, it is lack of foundation rather. Those why say that the panic was caused by the floating of non-paying securities are right, so far as they go, but they might have added that a large mass of securities must necessarily be non-paying, and it is the principles which make them non-paying, coupled with the attempt to make them pay, which are the real cause of the panic.

But some one remarks, if that were true we would

have financial panic all of the time. Not necessarily. There is a little attachment to the steam engine, called the safety valve. While that is in order the pressure on the boiler is not likely to become dangerous, unless some unusual freak is developed. Liquidation is the safety valve of the business engine. Statisticians say that more than ninety per cent of all business men fail. These failures occur every day, every week, every month every year. Those who fail are the persons who have failed to meet current necessary expenses of production, together with rent and interest. If they have failed to meet rent and interest, their property is taken from them, they bankrupt, formally, their slate is cleared and they are allowed to start new. The arrearages do not have to be made up. That is a very happy regulation, for the arrearages of everybody, under the present system, could not possibly be made up.

Again, panic is kept off for a time by using fixed capital as a pledge to borrow back the money paid in interest and rent so that it may be applied to the current necessary expenses of production and the business kept in the same hands. This process goes on until the capital remaining in the hands of the losing business man is no longer sufficient to secure credit. His obligations then come to a head, he fails. Starts are made about the same time, then failures come about the same time. It takes a time for debts to accumulate. An important failure leads to others being watched, pressed and driven to the wall. We have financial panics.

To make this point still plainer, let us take an illustration. We will suppose that a farmer owns fifty acres of land and on that land raises wheat. He sells his crop for six hundred dollars. Of that amount seed cost him forty-five dollars, taxes twenty-five dollars, insurance five dollars, the feed of team, one hundred dollars, deterioration of fences, buildings, machinery and stock, forty dollars, the threshing of grain, fifty dollars, and harvest help with board for the same, forty dollars. Let us suppose that the other two-hundred and ninety-five dollars is required by the farmer and his family to live on, and to educate his children. His necessary expenses, the necessary expenses of continued production, in his case takes all of the gross product and he can pay no rent and no interest. He saves nothing. Now, if he and his family could live on two hundred and fifty dollars per year he could save forty-five dollars in a twelvemonth or he might apply it to rent or interest without showing any signs of business failure or becoming a less effective factor in production. But let us suppose that the farmer can save nothing (which is but too grimly true). Let us suppose that every dollar of his gross product is required to pay necessary expenses of production, including his living and that of his family. Working in that condition, he is out of debt and owns his farm, so that neither rent nor interest enter into the calculation. Now let him lose one crop and be obliged to borrow six hundred dollars at seven per cent per annum. Immediately he must fail by forty-two dollars per year to meet his obligations. He must go on borrowing an additional sum of forty-two dollars per year, and when that with its interest aggregates more than his farm will stand as security for, he goes into bankruptcy.

The nation's industries taken on an average are analogous to this farmer (which, by the way, is an actuality and not a creature of imagination). They have an average interest charge of three and one-quarter billions to meet yearly and but a billion and one-half, after paying other expenses, to meet it with. These industries borrow while they can give security for the loans, then they fail, and we have a panic.

I do not mean that the industries of the country fail as a whole. They do not, for the country is not in business as a whole. But enough of the individuals engaged in industry in the country fail, so that their failure produces a panic and financial depression. For instance, A, B, C, D and E represent the individuals engaged in industrial pursuits in the United States. A and B gain three and one-quarter billions per year, while C, D and E lack a billion and three-quarters of making ends meet. The consequence is that C, D and E borrow from A and B a billion and three-quarters per year while the property of the former remains sufficient to meet the loan, and when that condition no longer exists, or confidence is destroyed, C, D and E file into bankruptcy and A and B possess themselves of the goods and chattles of their former business confreres.   C, D and E, in nine cases out of ten, bankrupt by trying to pay fixed charges greater than the productivity of the business warrants; by trying to make good the fabulous increase of the barren dollar and the barren wealth which it stands for.   There is nothing occult about it.   It is a natural consequence.   Periodical panic and wholesale bankruptcy are the legitimate result of trying to apply false principles to industry.

It is marvelous that bearded sages should gaze in mysterious awe at financial tornadoes which sweep over the interest-collecting world with the regularity of the tides, and ascribe them to this or that insignificant local cause, or shake their heads and say that they are necessary, but why they come no one knows, when conditions which must produce panic and bankruptcy are a necessary concomitant of the foundation principles of our financial and industrial system.

We assume in our industrial institutions, that wealth contains a natural principle or geometrical increase. We build our financial and industrial systems on that assumption.   We lend, borrow, trade, manufacture, mine, railroad, farm, upon that principle.   We find that for the majority of the people of the nation, it leads to disaster and distress. If we look a little further, we will find the principle absolutely false and utterly absurd.   Yet it works well for some of the people, the people who assume that as the elect they have a right to live by the labor of others, and because they say it is well all others believe it so.

CHAPTER XIV.

"The carriage of Dives every day throws the dust of its glittering wheels o'er the tattered garments of Lazarus."

The extremely rapid accumulation of wealth in the hands of comparatively few, is a fact needing no proof. It is a matter of common observation. It may be legitimately deduced from the census figures that eighty per cent of the wealth of the United States belongs to one two-hundred-and-fortieth of the population, or a fraction of one per cent. Even after averaging results so as to make wealth appear as generally distributed as may be, conservative sociologists deduce from the census figures that seventy-one per cent of the wealth of the country belongs to nine per cent of the population. Exact estimates are impossible, but really not necessary.

Within the memory of men still young, the millionaire was a very unusual citizen. Now he has grown to thousands, and a few of the class count their wealth by hundreds of millions. On the other hand, the tenant farmer and the tenant occupant has grown with amazing rapidity. The number who are classed as wage-earners, because they are poor, has grown to include by far the greater portion of the population.

One naturally inquires, what is the reason for all this? Has the rich man become more useful to his fellow man, that he should claim a larger share of the produce of industry? Not so far as any one is aware of. The time of the wealthiest and those whose fortunes accumulates the most rapidly, is spent in traveling over the world, sight-seeing, yacht-racing, gambling, feasting and toadying after foreign nobility, more idle and useless than themselves.

When the millionaire comes home, he goes to work planning and building mansions for himself, building monuments for his own aggrandizement, or perhaps corrupting politics if he happens to have a taste in that direction. This is the millionaire of the second or later generations. His study is, in short, how he can spend the most on his own little, insignificant person, and although he toils not neither does he spin, not Solomon in all his glory fared like one of these.

This millionaire had an ancestor, a hard-headed, close-fisted, cunning, calculating fellow. He (the ancestor) learned early in life that no great fortune could ever be amassed in productive enterprise; that the productive power of the effort of one man is very limited in amount and at best can only give him a competency. He also learned that if one wishes to become immensely rich, he must do so by appropriating the result of others' toil. He learned that there are just two legal methods by which this may be accomplished: (1), by appropriating the land which is to be used in industry and charging laborers all that they will pay and still use the land; (2), by controlling the wealth which was to be used in production and charging for its use all that was left after paying rent, except merely enough for the laborer to live on. While levying these charges, to be sure, he speculated; i. e., made the rent and interest charges extraordinarily large wherever he could, by controlling, at critical moments, the land and the capital of the laborers. Thus this hard-headed business man spent his life in appropriating what belonged to others. He laid a foundation by which his progeny might with greater ease continue like operations for all time to come. He was not particular about methods, if within the law. Wrecked properties and violated trusts too often strew the way of the millionaire. The enormously rich are not rich by remuneration for their services to fellows, but by taking all that they can get by any means at hand, and they find rent and interest the most convenient means.

According to estimates made above, the landlord and money-lending class save about two and one-half billions per year. One and one-half billion of this appears as an addition to national wealth, and another billion is lent back to business undertakers to tide them over to the day of reckoning. Thus this class would save about twenty-five billions in a decade and fifty billions in twenty years. This fully accounts for the increase of wealth in the hands of the very wealthy.

With a surplus each year to re-invest and an enormous interest-bearing capital, capable of absorbing all possible production, a caste of wealth must soon be formed, almost absolutely secure in the possession of their property. We must have a stable aristocracy founded on wealth. That class must in time have absolute control, as it will own all of the wealth. Interest will outrun the best inventive genius. The more wealthy this class becomes, the greater the number of persons who will be taken from productive occupations and retained by the wealthy to attend to personal wants, and the heavier will become the burden on the actual producers. Under the interest system the extremes will ever become more marked. But, to be more specific.

Following a convenient classification by another writer, we may divide the business community into two classes or groups: the financial class, who engage their talents in making gold and silver breed, and the industrial class, who perform all the work of production. All who lend on interest or use their wealth in any way to secure that which is produced by another, without parting with any of that wealth in return, are so far members of the financial class. All who produce or add to the sum total of national wealth are so far of the industrial class. The financial class lives by collecting rents and interest or incomes founded on these charges. The industrial class pays all rent and interest, as well as all other charges earned or unearned. Rent and interest are usually secured by liens on property, and must, usually, be paid. Hence the financial group, except in a small percentage of cases, receives its share of the product, and if any default is made, it is in the necessary expenses of production, due to members of the industrial group. Default can be made only in the payment of obligations between members of the industrial class. That is, members of the industrial class must bear all loss.

Thus, while each year sees thousands of the members of the industrial group go by the board and turn their effects over to the financial group to satisfy fixed charges, the members of the financial group lose little and grow constantly richer. While industry pays, they receive their increase; when industry fails, they receive both increase and principal.

There are certain beasts of prey which fatten by the misfortunes of their brother animals. There is a class of non-combatants who follow in the wake of fighting men and fatten on the spoils of the fray. They keep out of danger, skulking in the rear until the battle is fought, and when others are wrapped up in alleviating the sufferings of the unfortunate, they swoop down upon the wreck-strewn field and gather up the spoil. That is the position of a large number of the camp followers of the peace army, and whether the hosts meet with success or reverses, whether there is panic or prosperity, they reap a harvest. They are sure to avoid danger. During prosperity, they sow the seeds of disaster for others. When disaster ripens they get the corn, the toilers the tares. Interest is the winnower.

The rent and interest charges of three and one-half billions are about two billions more than can be met annually after paying necessary expenses of production, but, as rent and interest must be met, the members of the industrial group who fail to otherwise meet them are forced to bankruptcy or to borrow back a portion of the rent and interest fund in order to meet current expenses. For the wealth which has passed to the financial group is the only fund to borrow from except the half a million or so saved from the wages of labor and superintendence. The financial group lends only on

security, and the wealth necessary each year to pay
current expenses can be secured by the delinquents in
the industrial group only by hypothecating each year
an additional amount of their fixed capital to the members of the former group. It amounts, really, to an assigning a portion of the fixed capital of the industrial
group. Three-fourths of the net yearly increase of national wealth accumulates on the hands of the financial
group and, before it can be used in production, must
be returned to the industrial group. A half million
is saved in the industrial group, and if this is lent it
makes its possessors just so far members of the financial group, and begins a transfer of property from the
less fortunate to the more successful members of the
group of workers. In this way about two and one-half
billions yearly of fixed capital passes from toilers,
who constitute the members of the industrial group, to
the money lenders, or financiers, as they are pompously
styled. This is sufficient fully to account for the rapid
accumulation of wealth in the hands of the few and the
consequent impoverishment of the many. The change
is much more rapid than would be indicated by the
absorbing by the financial group of all the yearly increase of national wealth.

This assigning of fixed capital has but one result:
final bankruptcy. It must constantly cripple the business of the undertaker and cut down his income. If
he could not meet smaller expenses with a larger income, he cannot be expected to meet larger expenses
with a smaller income. He finally goes to the wall and
the financial group gets all of his property. This will
occur as soon as he fails to meet obligations or obligations become so large as to make the security doubtful,
and will, of course, make congestion of wealth more
rapid.

While it is not probable that those who receive incomes from rent and interest, spend from those incomes
more than a billion annually, such income amounts
to three and one-half times that sum. But about a
billion and one-half of this appears yearly in the increment of national wealth and the remainder must repre-

sent what is re-loaned to the unsuccessful to pay
current expenses of production. This sum gradually
gives the financial group control of the fixed capital
already in existence, as well as most of the increase.
The financial group thus becomes rich more rapidly
than the nation at large; and national increase in
wealth may not mean prosperity of the producing
masses. That is the fact. The financial group can be
prosperous, while they exact tribute from the masses
as large as business will bear.

To be sure, the classes are not rigidly fixed. A few
drop, each year, from the financial group into the industrial group, a few go from the industrial to the financial; but that neither makes richer those who are left
in the one nor poorer those who remain in the other.
Many individuals are members of both groups, but as
members of the financial group they get their incomes
from the industrials other than themselves, and thus
their double relation makes no easier the lot of the industrials who are not also financiers, but the opposite.
They do make it appear that they have earned their
wea th as industrials, and make it difficult physically
to separate the non-producing from the producing
classes.

England is sometimes cited as an instance of de-cen-
tralization of wealth. It is held that in England wealth
is now more generally distributed than ever before, and
this is supposed to prove that there is no danger of undue centralization of wealth. But conditions in England prove nothing as to other nations. England is
the loan-shop of the world, and her interest tribute is
levied on the whole world beside. Three-fourths of
the citizens of England may become rich by interest-
taking and yet not receive a dollar from a citizen of the
island. English wealth is produced abroad. Even
though the facts are correct, the conclusion of the
apologist is not warranted.

But, it is said, large fortunes are not made by interest-taking, but are amassed by speculation. The principle is exactly the same. It is the application of the
idea of taking all that one can extort by any means by

which it can be extorted, on the assumption that some-how his acts have created what he takes. The rate is larger in speculation, but the plunger loses so often, the business is so uncertain, that it does not begin to be as profitable as interest-taking. When great for-tunes are amassed, speculation is discarded altogether and interest-taking is used as a means of perpetuating the fortune. Then interest-taking, and a currency suited to that purpose, are the instruments which make speculation possible. It is by manipulating currency through the power of interest-taking that prices are speculatively affected and all speculative business car-ried on. Destroying interest-taking would destroy spec-ulation and thus do the country a service of inestimable value. The business of the speculator is appropriating, not producing, and he and his are a dead weight on the shoulders of the productive toiler. He gives absolutely no return for what he gets.

CHAPTER XV.

"The poor ye have always with you."

ON the other hand, we have abject poverty on the part of a large percentage of the producing masses. This is a necessary result of one class being immensely wealthy. They are immensely wealthy because they take what rightfully belongs to others, and these others are poor because their substance is taken from them and no return given for it.

As we have seen above, the taker of rent and interest gets all that he bargains for, and he bargains for enough to make him a millionaire and his victim a pauper. A man with one hundred thousand dollars lent on farm mortgages, can spend the average income of seventeen farmers and at the same time save the average income of ten farmers. Thus he may spend each year the salary of one of our congressmen and save sufficient to die worth two hundred thousand, so that his son may save twice as much and spend twice as much as he, without ever doing the slightest service to any one else on earth. Thus his posterity may become richer and more extravagant. The further they become removed from those who have done something useful, the more they can collect for idleness. Put at interest at five per cent, one hundred thousand dollars will double itself in fourteen years; at six per cent it will double in twelve years; at seven per cent in ten years; at eight per cent in nine years. This is much more rapidly than fixed capital can accumulate, and an attempt to carry out such accumulation compels a large percentage of the population to subsist on short rations.

As was shown above, there is a limited gross income to be divided between the landlord, capitalist and laborer. If any one gets more than his share the others must get less. But how very limited this income is, but few realize. If every cent that is left after repairing the annual deterioration of wealth, were given to the toiler, he would receive less than five hundred dollars per year. If the product were distributed equally between the inhabitants, men, women and children, each would have but about one hundred and eighty-five dollars to live on. Toilers cannot afford to share their meager income with idlers. The only way to increase wages and do away with abject poverty among producers, is to turn the rent and interest charges into the wage fund. If non-producers get three and one-half billions of dollars per annum, producers get just that much less than their share. When we consider that the whole product, besides enough to keep up capital, is barely sufficient to keep the workers, it is not difficult to see why some persons want.

I am aware that these facts are used by a few weak-minded individuals to support the taking of rent and interest. They tell us, with sublime simplicity, that this charge does not amount to enough to make any difference in the wages of the laborer, and show how little he would get if he were to get it all. It would increase his wages by at least fifty per cent.

Wealth is produced by labor alone, and the only way to increase the product and consequently the wages or share of the producers, is to make non-producers work productively. This can be done only by depriving them of their ability to appropriate a portion of the wealth produced by others. For, while one can live better by idleness, he will not produce.

If present non-producers became producers, several consequences would follow. The family of non-producers, who now spend one hundred thousand dollars annually, necessarily make one hundred toilers or more non-producers, in so far as other toilers are concerned. All that a non-producer or any of his servants consumes is consumed non-productively, so far as the toiler is concerned.

The horses which the non-producer uses, the man who cares for these horses, the man who produces feed for these horses, the man who feeds the man who produces feed for these horses, are all non-productive workers, in their relation to producing classes. What they do in no way aids any one who toils. For all this effort, all the product represented by the one hundred thousand dollars consumed yearly, is devoted to one object and but one alone: supplying the wants of one who does nothing to supply the wants of others. Industrially, it is a total loss. So far as toilers are concerned, except those actually working for the idler, it may as well have been burned on the spot on which it was produced, the only difference being that the loss would fall on different individuals among the toiling masses.

If that one hundred thousand dollars were spent by producers instead of non-producers, it would make.the former just that much better off, for every dollar would command a dollar in some other sort of service. The labor which that fortune now hires for the non-producer would be engaged in creating things to supply the wants of the producer. The coachman might be making shoes for toiling feet; the butler planting corn; the groom raising beef; the farmers who supplied the larder and stables supplying the toilers with bread. The non-producer, turned into a useful citizen, might be at the forge or lathe or counter, helping to increase the stock of wealth. At worst he could but go to the asylum for the feeble and could be supported at a trifling percentage of his present living.

The painter of daubs for the non-producing patron, might be applying a more useful brush to a house or barn to preserve it from decay; or if he was, perchance, an artist of real talent, he might be creating works of art for the edification of those who formerly furnished the wealth to pay for the pictures which the non-producer haughtily called his own.

To illustrate: let us suppose that five men engage in five branches of trade; one producing bread and drink, one meat, one clothing, one tools and machinery, one carrying these things from each to each, and a sixth

partner laboring to supply the æsthetic and intellectual wants of the community. Each produces what the
other wants, and each adds to the wealth of all. Now
let an outsider come who controls the land and the
capital which these men require, and compel each toiler
to give him a portion of what that toiler produces, as
well as to spend a portion of each day producing something which none of the toilers use, but which the landlord and capitalist wants for himself. Of course the
latter dignitary does not deign to produce anything. It
is enough for him to attend to his "business;" i. e.,
hatch schemes for taking a portion of what is produced
by those who work. It is evident that the time consumed in producing what is given to the capitalist is
time thrown away, so far as the toilers are concerned,
for some one else gets the benefit of that labor and returns nothing for it. If that time were devoted to producing what the toilers used it would make them that
much better off. It is a case exactly parallel to that
of the toilers of the nation. By destroying unearned
incomes, we must make all toilers, and make the effort
of each add to the wealth of all. This is the only way
to relieve distress. Our failure to do this fully accounts
for the abject poverty of many of those who toil.

CHAPTER XVI.

"It is doubtful whether improved machinery has lightened the burden of any toiling human."

THE invention of labor saving machinery. Improved machinery has failed to ameliorate the condition of the toiler in a degree at all commensurate with the increased producing power thus given to the laborer. Thorold Rogers,* a recognized authority on this subject, states that the labor of Englishmen in obtaining the necessaries of life was as effective four and one-half centuries ago as it is to-day. Even more so. Yet there is scarcely an occupation at which a laborer cannot accomplish many times as much in the production of wealth as he could even a century ago. What becomes of the surplus? As Mr. Rogers shows, the laborer gets but a small part of it in wages. Government, to be sure, is a little more expensive, but that does not begin to account for it. There is but one other place to which it can go: as unearned income to the landlord and the usurer.

Instance the one occupation of agriculturist. With the gang plow, the seeder, the horse rake, the sulky cultivator, the horse-fork, the self-binder and the power thresher, one man can accomplish the former work of at least half a dozen. Has he half a dozen times as much wealth at his disposal each year? Not at all. Unless he lives well nigh as primitively as he did three-

*Work and Wages—539-541 et seq.

108

quarters of a century ago, he cannot make ends meet.
People say he is extravagant, his family want comforts
which their grandmothers did not have, hence he is not
prosperous. These critics scarcely realize that the
prodigal luxury of the farmer is held within the bounds
of three hundred dollars per year, while the critics
groan because they are limited to from twenty to forty
times that amount. And the farmer is not an exception in this respect. The man who by use of the machine makes half a dozen pairs of shoes where one could
be made a century ago, or ten yards of cloth to one
made by his father, or ten tons of iron as readily as one
could formerly be made; or carries a ton of produce ten
miles as cheaply as it could formerly be carried a mile,
receives little more real wages than his fathers did. To
be sure, the laborer of to-day has things which even the
wealthy could not have a few years ago, but for the
substantial necessities of life he is but little better off.
His environment compels him to live on a different
scale from that of the past. It makes necessities to-
day of the luxuries of former days, so that what he
must have presses as closely as ever to the limit of his
income. And as compared with the wealthy, the laborer
of to-day is infinitely poorer than his laboring ancestors
were. The effect of machinery has so far been the piling
up of large fortunes, rather than lightening the burdens
of toil.

It is all due to the principle of interest-taking. The
march of machinery has not increased net production
in a geometrical ratio. Interest demands the geometric increase of product to keep pace with its demands.
Interest was and is charged, as we have seen, on nearly
all capital used in production, directly or indirectly.
Interest-takers, therefore, control all of the product
and claim all that is left after giving the majority of
laborers a mere living.

We will suppose that a man is able by his toil with
the implements of three-quarters of a century ago, to
secure a living for himself and his family. Such was
the fact. He owned his farm or shop and retained
nearly the whole value of his product. The inventor

came, with his improved processes. A business man adopted it and loaned it to the farmer or mechanic, with the understanding that the lender should have the lion's share of the product. The contract is carried out. The lion's share of the product goes to the lender in interest and is re-lent to bear interest; the charges which the mechanic has to meet multiply and he finds that he still has left barely enough to support his family. But the lender has also enough to support his family, and to support it on a scale of which the mechanic never dreamed. Thus it went with every invention. It was controlled as capital, lent, and the increased effectiveness went largely to idlers, under the laws of rent and interest.

The inventors even, as a rule, get little of the benefit of the fecundity of their ideas in the world of industry. They, as a class, die poor. It is the man with the capital which is assumed to grow, who is lord of creation and for whom all other men toil. He is the interest baron, the collector of increase on decaying wealth!

Then, in the manufacturing industries, the increased use of machinery tends to throw the laborer out of employment and fill the labor market with men eagerly competing for places. The capitalist can run his business with fewer hands. The competition of wage-earners becomes more fierce. Those who can secure places do not dare to ask for more than they were receiving when laboring with poorer mechanical appliances, for there are many to take their places at such a wage. As a result the capitalist gets the whole benefit of the increased effectiveness of the machinery. It is paid in interest, rent and capital profits, which differ in no way from interest. The effectiveness of invention, under the system of interest-taking, accrues to him who controls the material of which the machine is made. While interest and capital profits are collected, invention must have the immediate effect of decreasing wages, and making fewer bear the burdens of productive toil.

CHAPTER XVII.

What one relinquishes, not what one can extort, is the measure of his service.

NON-PRODUCERS have much larger incomes than producers. In fact, one's income is often in inverse ratio to the service which he does his fellow men. This fact cannot be accounted for on the theory that every man should be rewarded according to his services, or that each should have what he produces and that alone. It must be accounted for in some other way.

As for tangible material production or value of services, seven hundred dollars per year is the utmost average limit to human power, and as one-third of this is eaten up in repairing the natural decay of capital, less than five hundred dollars per year is left for the wage of the average toiler.* Men may sometime become morepotent producers, they are becoming so each day, but the above is the limit reached with the present quality of land and machinery. There is a difference in indi-

*This, to be sure, is but an average at best, but it is not far from the mark. There are probably twenty five millions of persons, including women who take care of homes and children, in gainful occupations. The yearly gross product is not more than fifteen billions, and this would be a product of but six hundred dollars for each worker. It would take a product of twenty billions a year produced by twenty million toilers to bring individual product up to one thousand per year, the figure adopted by Mr Atkinson. On the other hand, there are many returned in census reports as being engaged in gainful occupation, who never add a dollar's worth to the wealth of the community, and there is no way of re-estimating the number of this class. Hence an estimate of average *per capita* product is more or less a guess, but I think I have put the figure high enough. Even at one thousand dollars, the *per capita* yearly product could not account for yearly salaries.

viduals. Counting extremes, there is a wide difference
in the ability of individuals to produce. But that difference does not begin to account for the difference in
wages. The largest giant is scarcely twice as tall as
the smallest dwarf. Just so the most prolific worker
can scarcely produce twice as much as the most ineffect-
ual producer, in full health and vigor. Yet the yearly
income in one case is often one hundred times that of
the other, or even much greater.   •

The wages of superintendents should be only as much
greater than those of the producing rank and file as the
ability of the superintendent could increase the
product of those whose labor he is directing, beyond
the limit which could be reached under the direction of
any person within the rank and file of the producers.
Experience teaches us that talents do not differ greatly
in the same class, and that in any large body of workers
there may be found several who could do quite as well
as superintendent as the person who actually holds the
position. This is verified every day by promotions from
the rank and file to the superintendence and management in all branches of affairs, public and private.

Then, either on a competitive basis or as a return for
actual services, the wages of superintendents should
be little if any above the wages of the best workers
whom they direct, and not very much above the average. The facts are largely in accordance with this where
the superintendent is a mere hired foreman and does
not include in his remuneration, capital profits of some
sort.

Then the pay of the actual toiler or the toiler of superintendence, based on tangible product, cannot be
rightly much above the limit of the tangible average
production of one individual. This would hold the
maximum salary down to about one hundred dollars
per month.

But with the principle admitted that one has a right
to take as much as he can get, and to get as much as
he can by any means at his disposal, whether it has
been produced by himself or others, there is no limit to
what one may take as income, or, more accurately, it

is limited only by what he has taken already. The capitalist, with one hundred thousand dollars, can levy tribute on one hundred laborers, and take from eighty to one hundred dollars per year from each, giving him an income princely as compared with the incomes for these laborers. He can afford to pay a non-producing lawyer two thousand dollars per year to do this for him, so that the man with capital need not turn his hand, and still he will have a goodly net income. The capitalist with a million can pay ten times as much to a parasite of the law to financier for him and still live like a prince on the tribute exacted from toil. Under the law of interest, the wages of idleness may be any sum falling short of the gross product minus the cost of an indifferent living to the majority of producers, while the wages of toil is rigidly fixed at a most moderate allowance.

But, say the apologists, these men are paid for brain work. So is the cracksman who loots their safes, but it is brain work of the sort which should not be encouraged. It is the human intellect directed to the problem: "How much can I take from others without giving aught in return?" Mental labor should have its reward. The man who invents a useful machine; the man who subdues a natural force to the use of man; the man who stirs the souls of men with noble thoughts, inspires the living glories of the magic canvas, or the glowing life of the chiseled marble, is a public benefactor, deserving of substantial reward. But does the capitalist claim remuneration for any of these acts Not at all. His remuneration comes from the exercise of that stealthy, cat-like cunning which in the days of force enabled primitive man to steal upon his fellows, grasp a momentary advantage, and carry off property after wasting life. He takes because his wealth gives him the power to do so.

That the brain work of the benefactors of the race should be remunerated is no reason why the able lawyer, who uses his God-given intellect to pervert the laws which he has sworn to support, in order that a great corporation or a petty thief may take from others

what their toil has produced, should be looked upon
as a benefactor or receive, from the very people whom
he labored to injure, a princely remuneration for his
treasonable iniquity.

I am discussing this question from the standpoint of
the real producer, from the standpoint of the greatest
good to the greatest number. Ninety per cent of the
people have no interest in supporting the pretensions
of the other ten per cent, who assume a superiority to
their fellow beings and think that somehow they (the
wealthy) were divinely commissioned to enjoy all of
the good things of this life at whatever cost to the remainder. The standpoint of the producing toiler is the
only standpoint from which any honest economist can
look at the subject. Either men have equal rights to
what they produce, or they have not. If they have not
some men must be naturally or divinely delegated to
rule over their fellows and appropriate what they
choose. The appropriator, in a state like this, must
be the judge, for there are none other. Might is right
and cunning might. There is no limit to extortion,
ethics is but a myth and economics the mutterings of
fools. The people must take thankfully what their lords
and masters give them. There is no middle ground.
One must either accept the foregoing or admit that
men have equal rights in all things, and base his economic philosophy on this principle. On the principles
of equal rights of man, I hold that because intellect
applied to the aid of toilers should be rewarded, is no
reason why intellect applied to despoiling toilers should
be-rewarded.

The great money-getter may work hard, have anxious
nights and sleepless days, but if that toil and anxiety
is directed to wrecking a railroad that he may gather a
fortune from the ruins, his toil should receive the meed
of the criminal, not the reward of a productive laborer.
Is his toil directed to getting wealth for himself at the
expense of others, or for others to whom it does not belong that he may share the plunder, or is it directed to
wresting from nature wealth for the benefit of all, himself included? On the answer to this question depends

whether the mental or physical effort of this or that particular person is rightfully entitled to remuneration. And it applies to the lawyer or the bank president or the railway financier as well as to the cracksman or sneak thief.

As to the relative amount of remuneration which mental and physical effort should control, where both are productively employed, the present practice seems to be based on dubious grounds, to say the least. We assume that mental effort is worthy of greater remuneration than is physical, but if required to give a valid reason for such assumption, we would not find our way at all easy. Mental ability is its own reward to a greater extent than is physical, and mental effort cannot claim remuneration on the ground of being at all exceptional. A railroad president collects fifty thousand dollars per year for his services, because he controls stock enough to vote him that salary, not because his services are worth that much to the community which the railway serves. They may be positively detrimental. It is unnecessary to mention names or instances. Look at the railway history of the last thirty years. And as for exceptional talent, it is no exaggeration to say that there are ten thousand men in the United States to-day, each drawing a salary under five thousand per year, who would make railway presidents as able and reliable in every respect as any now drawing fifteen to fifty thousand dollars for such services. Judging from results, railway officials, notwithstanding the princely salaries, have been criminally negligent of their trusts, or else criminally inefficient. Railroads make a poorer showing than any other class of business enterprises, having had one-third of the whole mileage in the hands of receivers at once.

A lawyer gets one hundred thousand dollars for a case, and his whole effort has been toward oppressing the toilers from the product of whose toil he is remunerated. The large fee is usually for unscrupulousness. There are a thousand others, whose yearly incomes are not one-twentieth of that sum, as learned in the law as he. Indeed, it is an open question whether if every lawyer in the United States were disbarred to-day and turned to useful occupations, and justice were administered on man's sense of equity and a common sense interpretation of the law, the cause of real justice would not be furthered, and the community be in all respects better off. We would certainly have a less complicated and more comprehensible system of law. Any man with a fair English education and a degree of common sense could draft laws more intelligible to the average citizen, who is supposed to understand them, and infinitely more terse than the product of the lawyer. Without the quibbling lawyer these laws would discover little ambiguity, common sense could easily find the intention and interpret in that light. The effort of the lawyer seems to be to make laws, as lawmaker, which he, as a lawyer, can pervert for the benefit of his client, whether that client be seeking that which of right belongs to him, or seeking to despoil another. Lawyers have meshed the laws into such a Chinese puzzle that no one pretends to know a tithe of the law of the land, although there is an agreeable fiction that no one is ignorant of the law.

I do not mean to insinuate that lawyers as a class are naturally less honest or more obtuse than other men, but I do affirm that most of them are striking instances of good talent perverted to unholy purposes. They are a necessary product of a vicious industrial system and thrive on the reprisal element of modern commerce. From the standpoint of the toiler they are certainly entitled to no more remuneration than that claimed by the average worker.

The doctor with enormous income is as often the charlatan whose profession of empirics and experiment applied to questions of life and death he uses as a means of extorting money, as he is an honest, earnest worker, trying to learn the laws of health and to impress them on his patient. And the learned doctors who get less than five thousand per year are twice as numerous as those whose eye for extortion and dishonesty gives them greater incomes. Strip medicine of its pretense, of its juggling with the sacred right of life. Let doctors tell the truth and throw away drugs which the most learned will not positively assert to be of benefit to the sick, and the twenty thousand per year practices would be so few that they might be ignored, and the number of doctors would be reduced by one-half, with infinite benefit to the health and pockets of the community. On the basis of a return for what it receives, no community can afford to pay a doctor twenty thousand per year, unless it be for his usefulness in getting rid of undesirable characters, and then it should induce him to take his own medicine.

Neither medical nor legal fees should be so high that the best talent, when needed, may not be at the command of the humblest, and this is clearly impossible where the services of a professional man for a few hours' work is greater than a toiler's salary for a year. Highly paid lawyers and doctors simply share the unearned incomes wrung by capitalists from industrial workers through interest-taking, and if these unearned incomes were done away with the lawyer or the doctor would be obliged to give his services for what they are worth. I fear that some of them would have very slight remuneration.

Literary men hardly ever get very unusual salaries. Where they make an exceptional amount it is usually as business men. It would be difficult to measure their just remuneration. If, as is often assumed, they had a monopoly of the thought which they express, their remuneration should be very large, indeed. But the fact is, that the burning thoughts which they express so well were the common property of a thousand brains before their accredited authors ever put pen to paper. If they had not expressed them, others would. Thoughts are a product of community civilization as truly as land values are. Shakespeare's age had a hundred lesser Shakespeares who might have been greater had Shakespeare never been. Has the reader never had a thought which he imagined all his own until on looking over printed pages he found that thought expressed better than he could have expressed it, and already given to the world? I think it is the experience of every one

who thinks at all. Indeed, literary glory depends more on
the manner of expressing than on the ability to conjure
up great thoughts.. Back of the whole race of throbbing,
sentient beings there seems a great volume of knowledge which kindly nature turns over .page by page before the eyes of all who care to read. It can be read
only when the page is turned, then it is common property. There is some great world spirit sitting on that
throne of knowledge which will allow no one to be a
miser with his thoughts. If you would keep them, use
them first. As great thoughts depend on no one person
in particular, no one can claim any especial remuneration for great thoughts.

The enormous prices paid for works of art are a relic
of the barbaric spirit, still strong in the child, to want
that which others cannot have, instead of enjoying it
all the more because others enjoy it. It is indulged by
the possession of unearned incomes founded on interest,
and would be destroyed with the destruction of these
interests. Those thousands of dollars are not for the
artistic ability displayed, but to gratify a barbaric
pride. A people can afford nothing which all may not
enjoy.

All great devices are invented by degrees, so much
so that they are the common property of several before
they ever see the light. Like the infant learning to
walk, man's steps.into the field of knowledge are feeble
and tottering He cannot leave the bench of present
knowledge, but must drag it with him as his support
in fields unknown. If Fulton had not invented the
steamboat, some one else would. Stephenson did but
a little more than some others toward perfecting what
was initiated by a Greek or Egyptian, who can say
which Greek? There was a world of knowledge upon
the history of civilization in one of the railway exhibits
at the World's Fair, and it said in thunder tones, how
insignificant is the influence of any one man! Edison
has patented what a hundred others were toiling with
as well as he. Those who are known as the inventors of
this or that machine are entitled to no very especial
remuneration.

It is the same with great discoveries in science. The observations of the Chaldean shepherds were necessary to the truths enunciated by Laplace.

The very large pay of the so-called histrionic artist or concert singer is founded largely on the fashionable extravagance of those in control of unearned incomes. It is the same principle as that which governs the price of works of art.

Who can say who invented printing? Who wrote the books of Moses or the Zend Avesta? Who wrote the poems of Hesiod? Who were the pioneers in Anglo-Saxon literature? Who really gave the laws of Solon? Who first discovered America? Who furnished the ideas for Hamlet or the Divine Comedy? Who will answer? The more the person questioned knows about these things, the less likely he is to be dogmatic. The great individual is but a bubble on the surface of the mighty current of humanity, a bubble sent up by a trifling fortuity. He has as little to do with the headlong course and sweeping might of that current as the chip on the broad Mississippi's breast has to do with directing the course of the great father of waters. He simply shows whither it sweeps. While none are overmastering, none can, on the principle of reward for services, claim an overmastering prize. To the swiftest is the race, but the rules of the course should be the greatest good to the greatest number. He who reaches the goal a second ahead of a thousand others, should not get a prize as large as the thousand, especially when his gain is due to a better start.

While talents differ widely, from the highest to the lowest, as the giant towers above the dwarf, no one has talents far above all of his fellow men. The giant touches shoulders with other giants. No man is then entitled to any very exceptional remuneration on the score of his services to other men. No man ever earned fifty thousand per year, not to speak of saving from twenty to fifty times that amount. There never was a man on earth, and never will be, whose services might not have been dispensed with without making the earth perceptibly poorer in any respect; and some of the

most valuable have received scarcely any material remuneration. The vast difference in the remuneration
of men depends on the power of some to control the
results of others' toil, and that power depends on rent
and interest. Men have long since abandoned the idea
of getting rich otherwise than by appropriating what
rightfully belongs to others.

Ability is the great factor in production and ability
should receive the reward, wisely remarks the apologist.
What ability? Ability to appropriate or to produce?
As we have seen, ability does not differ widely in
tangible material production. As we have seen, inventions of great effectiveness are dependent on no one
man. As we have seen, talent is not rare; what any
one man has accomplished, thousands of others might
and would have accomplished. Otherwise the place of
no important personage could be filled, but they are
filled by the hundred every day and the world does not
see the difference. But as we have seen, it is not inventive power, nor genius in production, nor great
merit of any sort which receives the material prizes.
It is low, stealthy cunning and unscrupulousness,
largely.

If ability to serve were the measure of reward many
men in high places would now be living very frugal
lives. The ability which is exhibited for an important
consideration is not rare. Any one of twenty states in
the Union could supply a president, cabinet, congress
and supreme court, who, with as much training, would
be as efficient as our present government force and take
no man who earned over five thousand dollars per year
either. The fact is that the highest salaried officers of
government or corporations are merely figure-heads so
far as the practical business is concerned. The president's labor is ended when he selects a cabinet. If it
is not, the country usually realizes that fact to its ever-
lasting sorrow. No matter how able, the president
cannot have knowledge of details in all branches, sufficient to enable him to make an intelligent decision on
the practical questions which come before him, and he
must take the judgment of some one whose efforts are

confined to a narrower field. The people decide the principle which is to be applied, so the chief executive is only a sort of button for under officers to touch in carrying on the affairs of government. The same is true of all officers of large corporations. They must depend for their action on the judgment of lawyers, promoters, confidential clerks and managers. And it is no reflection on the officer to say this. From his position he must necessarily be a figure-head, except, perhaps, in his judgment of the men about him, which he has every opportunity to know. Hence the high positions require less real ability than those considered less important. On the score of required ability the confidential clerk, attorney or promoter, the cabinet or assistant cabinet officer should have larger income than the head of the corporation or the government, but such is not the case.

Ability has nothing to do with remuneration; remuneration or income depends on one's power to appropriate and this depends on interest and rent-taking.

We often hear it said by apologists for plutocratic privileges, that this or that income does not come out of the people's pockets and the people have no cause for complaint about it. People, they say, have a right to do what they please with their own. This is sophistry of the rankest sort. People often have not an ethical right to do what they please with what is legally their own, for the reason that it is not ethically their own. And as for the people having no interest in the payment of big salaries to railway and other corporation officers, they have just as much interest in this as in the manner in which taxes are disbursed. Wealth does not create itself, then it is created by productive workers, every dollar's worth of it. Productive workers, then, have the right to demand that every dollar of it brings them a return. If a railroad pays exorbitant salaries or scatters passes broadcast, the patrons of that railway, ultimately the productive toilers, have to bear the extravagance in increased rates. If speculators parallel lines where commerce does not demand it, keeping up the extra line is a tax on producers. The average citizen has just as much interest

in seeing railways run economically as he has in seeing
the government administered with economy. In fact
every business in the country affects every other and
any extravagance in any is a loss to every toiling citizen.
In society, man is bound to his fellow man and has
obligations to fulfill to him whether he likes it or not.
If I am a boot manufacturer and pay a superintendent
an exorbitant salary, cutting the wages of my workers
in order that I may do so, I primarily wrong my employés and indirectly wrong every productive laborer
in the country. I attack the financial interests of all
toiling citizens. If my employés had what they earned
they would take more of the product of other concerns·
working in their class and make their wages better.

To be sure, when one has produced a thing it is his
own. He may then do what he likes with it without
affecting any one, but he must see that he has produced
it himself or given its producers an adequate return for
it before he can assume to do with it as he pleases. If
one wants to become irresponsible to his fellow man
let him go on a desert island where no being exists,
and then he is irresponsible only until some one else
comes. When another claimant to the island arrives,
some understanding as to joint possession must be en-
tered into and the men become immediately inter-re-
sponsible. And even as to wealth, it is a trust, no man
possessing a title other than for use, present or future.

CHAPTER XVIII.

WORK A BOON—Involuntary idleness—Extraordinary expenses incurred in times of financial distress—Why they seem to help the condition of the country.

The "curse of humanity" is eagerly sought for by thousands.

Why is work looked upon as a boon, and like Macbeth's power of prayer, he who has most need of employment is left without it? It is evidently because he has not the means of applying his labor to natural opportunities, or the natural opportunities themselves are withheld from him. While he remains part of civilized society, either will deprive him of his ability to work. Monopolize the wealth which has already been produced and you can control further production, dictating terms on which the wealth may be used or preventing its use at all. Wealth is monopolized by controlling the surplus, and this is made possible by rent and interest-taking. In times of panic men who control capital do not wish to risk it in industry. They hoard it and the laborers remain idle and starve. Interest and rent-taking bring about panic, and hence the distress of the laborer. If wealth remained in the hands of the toiler, he would certainly employ himself when he needed employment. Rent and interest take it from him.

When a business man is in financial trouble, he lops off every expense possible and lives closely until he gets out of his difficulty. He would not think of undertaking a new building or incurring any extraordinary expense, while pressed for money. Yet, at times of panic and depression, the most extravagant schemes of public improvement are undertaken, entailing great additional expense, and they actually help the condition of the country. This seeming paradox is explained by the fact that a portion of the hoarded capital which was withheld from laborers is released and the laborer is given an opportunity to work and create an effective demand for the wealth tied up by the hoarding of other wealth or its currency representative. The embargo of the rent and interest collector is raised. If wealth were not in the wrong hands, care would be taken to place it first in the most remunerative employment, and extraordinary improvements would naturally be left to prosperous times. While a few men have, through rent and interest-taking, power over the fixed capital of the country, they will hoard it when in danger and leave the bulk of laborers idle and starving, and all business paralyzed. Capital should be in the hands of those who must use it to earn their daily bread.

CHAPTER XIX.

Man versus Mammon.

War, the great destroyer of life and treasure, often ushers in new eras of national prosperity. It is certainly not the destruction of property which has benefited the nation, for that makes everybody poorer. It is merely because of its leveling propensities. The prosperity of a nation depends on the prosperity of her common people. A destructive war brings from their hiding places the hoards of the wealthy. It decentralizes wealth.

The immense destruction calls for immense production, so that all capital must be employed. The scarcity of laborers makes it necessary for capitalists to pay higher wages, i. e., give the laborer a greater part of the wealth which he produces. He therefore consumes more and is a better customer to other productive workers. Business is active, times good. The capitalist is obliged to disgorge and allow his wealth to be used in production lest he lose all. He cannot exact of toilers such heavy tribute.

History is full of testimony as to the salutary effect of wars on industrial prosperity. According to Ramsey, the common people were unusually prosperous during the Revolutionary war. We still hear laborers speak of the "good times" during the war of the Rebellion. The French of the revolutionary period held monarchical Europe back with one hand and at the same time gave a better living to the common people than they had known for centuries. They were prosperous because relieved of the tribute to the usurer and landlord.

It is a grim fact in the history of the world that unwarlike nations become peopled by races of slaves. The more robust the war spirit, other things being equal, the freer and more prosperous the nation. Instance a comparison between the nations of Europe and those of China' and the east. This fact is inexplicable except on one hypothesis, for war is in itself destructive, and the military rule inimical to liberty. But the explanation is that war prevents the accumulation of wealth in the hands of a caste, it is a great leveler and equalizer. It is in that way a heroic remedy for a terrible malady. Of course, I refer to wars which stir up the whole population.

'It is the wealth which is absolutely secure in the same hands which is dangerous to the liberties of a country. Show me a nation where revolution is unknown, and I will show you a nation of serfs. I can go further and show that revolution and serfdom usually appear in inverse ratio. This does not go to prove that revolution is in any way desirable, but that it is more so than the desperate malady which it is intended to cure. Wealth must be decentralized by some means. England, at the van of progress, is a country where the revolution is usually silent and seemingly infrequent. Her active spirit of colonization and the opening up of new fields of industry to her people, as well as the broad commercial spirit of her sons, have served to neutralize the tendency to crystallization in her economic circles and to preserve the industrial and political liberty of the masses. Shut England up by herself and let her inhabitants prey upon one another only, as the inhabitants of other nations are accustomed to do, and the country would be no exception to the proposition that some influence, disturbing .to vast centralization of wealth, is needed to preserve the liberties of a people. The opening of a new continent, of itself, brings about a revolution. In new countries, class-making must begin anew. The spirit of equality which is inspired by the reaction of the new nations upon the old, acts as a leveler, and puts off the plutocratic crisis.

CHAPTER XX.

A law-sanctioned wrong to the humblest citizen is a national wrong.

WHY is it that the accumulation of wealth has heretofore proved the destruction of the greatest nations? Up to a certain limit, wealth is power. There is no reason why it should enervate a man to supply his legitimate wants. Leisure and refinement should give man greater control over himself, teach him better how to live, to govern, to secure stability to the state. Yet we have Egypt and Assyria and Rome and Carthage and Greece to testify to the fact that great accumulations of wealth in the hands of a few of a country's citizens means moral, political and industrial decay. We may yet learn the lesson from a more modern tutor.

On the theory that each man is entitled to what he can extort, this fact of the destructive power of wealth is utterly inexplicable. It revolts our common sense as well as our sense of logic, to assume that what is right can lead to a radically evil result. If unlimited appropriation is right, the civilization founded on it should be righteous and enduring. If such has not been the result, then unlimited appropriation cannot be right. That is where the explanation lies.

The prosperity of a nation depends on the prosperity of all its people. If any considerable portion of the inhabitants of a nation are ground down, the nation itself will soon feel their misery. The nations of old, like the nations of to-day, permitted interest and rent-taking. The wealth of the nations accumulated in the hands of the few. The laborers of the nations became

127

poverty-stricken. Poverty, abject and continued, bred
vice, lawlessness, decay. Men became hungry, then
craven, then brutal, and lost all idea of justice. Like
the hungry wolf, the gnawing wants of the animal alone
were felt, and no method or means which would supply
that want was scorned. Men became slaves, women
strumpets.

On the other hand the means used in accumulating
enormous wealth drove from the soul of the wealthy all
idea of right. They began to doubt that there was any
such thing as right in the world. The idle, useless leisure which they obtained through their wealth oppressed
them and they plunged from one excess to another.
They had the means to gratify every whim. They had
no need to resist any self-indulgence. They soon became unable to resist it. Producers became worthless,
their masters more worthless. A state can no more be
upheld without citizens, than a great building can be
constructed of mud bricks. The state crumpled, or fell
a prey to stronger peoples whom wealth had not cor-
rupted.

While the causes remain, the effects are sure to follow. We are in this great republic following in the
footsteps of Roman luxury. We are undermining the
integrity of citizens by the extremes of poverty and
wealth, and if we keep on we shall pay the penalty.
Give to every man what belongs to him and none will
have wealth to grovel in, none will feel the imbruting
influences of poverty and want.

CHAPTER XXI.

"But a bold peasantry, its country's pride,
When once destroyed can never be supplied."

The yeoman of America, the independent farmer, is slowly passing. His toil is unprofitable as unceasing. He is dropping behind in the march of progress. He has become the jest of the narrow, ignorant, shallow minion of plutocracy, but a warning to all thinking people with the interest of their country at heart.

It is true that we have passed, to some extent, from an agricultural to a manufacturing population, but this does not account for the abandoned or dilapidated farm. There are millions more mouths to be fed. The demand for farm produce has not fallen off, yet farming is unprofitable. The margin of profits in agricultural pursuits has dwindled to zero, and the product of the farmer is not now sufficient to reward the toil of the producer. I have already given an instance, but I will make it more specific. On well known business principles, agriculture cannot long continue to pay much less in one portion of the country than in another. The business will soon be abandoned where it is least profitable. A typical instance in one section of the country is largely typical of the situation in the whole country. To simplify the matter, we will assume that the farmer owns his farm and works it with his own hands. A man with a good team can work a wheat farm of fifty acres, with the aid of an additional hand for about thirty days at harvest time. This will give employment to the man and team for the whole year, and

129

rather arduous employment it will be. A farm of fifty
acres, in Minnesota, for instance, within one hundred
and fifty miles of Minneapolis, the great flouring center, is worth, with buildings and improvements, about
two thousand dollars The team and machinery sufficient to enable one man to work that farm is worth,
or was during the decade from 1880 to 1890, about five
hundred dollars. The land and capital employed then
would be worth about twenty-five hundred dollars,
something less than half the value being land and half
wealth. The expense of working and maintaining that
farm for a year would be substantially as follows:
Seed, $45; feed for team, $100; deterioration of improvements, $40; taxes (local, state and school), $25;
threshing, $50; insurance, $5; hire of hand with board
(thirty days during harvest season), $40, making a total
of $305, besides the labor of the farmer.

The very best land with the very best cultivation
will not average, year after year, more than twenty
bushels to the acre, but I take this return, as it will
be equal to any crop the farmer may raise and in that
way represent the case of the mixed farmer also. The
census figures give an average yield in different years
of from about thirteen to less than fifteen bushels to
the acre. Wheat requires less labor than any other
crop. Placing the yield at one thousand bushels or
twenty bushels to the acre would give the farmer a
gross return of six hundred dollars, or a net income or
wage of two hundred and ninety-five dollars. This he
must live on and feed, clothe and educate his children.
And this is the wage of the prosperous farmer, the envy
of his fellow farmers.

If the farmer of the northwest could be sure of twenty
bushels of wheat to the acre and sixty cents per bushel
therefor, or a crop of mixed product equal to that, we
should hear little murmur of hard times. Yet what a
pittance is two hundred and ninety dollars per year
with which to raise a family, who are supposed to take
their place as intelligent American citizens and to direct
the destinies of the nation!

To be sure, the most common sort of farming is mixed

farming, where men, women and children engage in the toil, but the small farmer's income is scarcely ever more than is indicated above.

When you put the farmer, then, on a leased farm or require him to meet a mortgage upon his own, his living falls below the line of respectability. He and his family are required to try to exist on one hundred and fifty or two hundred dollars per year.

The American citizen will refuse to do this. He fails, goes to the cities and swells the crowd constantly surging against the line of starvation. The peasant farmer from Europe takes the farm, and our yeomanry is turned into a poverty-stricken peasantry with traditions of kings and slavery. The Americans who remain soon reach the same condition. There are but two ways left: either to fall to the scale of living of the European peasantry, or let the debt-burdened farm go.

The secret of the whole trouble lies in interest-taking

If the farmer is not paying interest on a farm mortgage, he is paying interest on the bonds of the railways which haul his wheat, and probably besides princely salaries, dividends and speculative profits. He is paying capital profits on a hundred things which he is obliged to use and capital profits and speculative charges on the grain he sells. Or rather, the consumer is paying these charges without giving the producing farmer any benefit.

Farm life is lonely enough without adding to it the gloom of want. There is no toil so exacting, no occupation has such a monotonous round. It is no wonder that the youth fly from it and that the aged become bent and hard and grasping and narrow, a beautiful subject for the plutocratic tool, too hair-brained to see the tragedy of what he caricatures. More than likely he has escaped from the dull and awful grind to turn traitor and spurn the stepping stones to his success, if he has any.

If the American farmer is not to pass to tenancy through the gate of foreclosure or bankruptcy, he must be relieved of direct or indirect interest charges.

The line which divides the tenant farmer of America

from the European peasant is fast being obliterated, and if America would secure the integrity and independence of a vast body of her citizens, she must see to it that the burden of the farmer is lightened. Garrison said that you could not enslave one human being without menacing the liberties of the world. You cannot degrade one human being without degrading the whole nation.

CHAPTER XXII.

Debts are never assets.

THE magnitude and rapidity of increase of the mortgage indebtedness of the country is little short of amazing. During the nine years preceding January 1, 1890, the mortgage debt of the country increased more than one hundred and fifty per cent, and the percentage of increase of mortgage indebtedness contracted in 1890, over the indebtedness contracted in 1880, was more than one hundred and forty per cent. In 1889 the mortgage indebtedness of the United States was six billions nineteen millions and one-half. At the rate of increase which obtained from 1880 to 1890, we would now be paying interest on a mortgage indebtedness of more than ten billions.

I am aware that a certain wise economist pronounces mortgage indebtedness a sign of prosperity. I would have more regard for his sincerity if I had less for his intellect. It would be difficult to convince a gentleman of Mr. Atkinson's experience as a hard-headed business man, that debts are assets, or that the larger the debts of a firm, relative to its assets, the better their financial standing. If he were to find by consultation with his "Bradstreet" that a certain firm last year had debts to the amount of but one hundred thousand dollars, and assets to the amount of five hundred thousand, and this year had debts to the amount of two hundred and fifty thousand, while their assets represented but five hundred and fifty thousand, he would say: "That firm has lost one hundred thousand dollars within twelve months; they will bear watching." The nation is in just that condition. Debts have increased much more rapidly proportionately than wealth.

Indeed, in Mr. Atkinson's own article, founded on the assumption that debts are assets, he moves heaven and earth to disprove the proposition which he lays down on starting out. If debts are a sign of prosperity, the prosperity of the American agriculturist should be measured by his indebtedness. Yet Mr. Atkinson asserts first that debts are a sign of prosperity, then labors to prove, and proves to his own satisfaction, that the debts of the farming community are relatively small; then, with amazing inconsequence, concludes that the farming community must be prosperous! One of the leaders of plutocratic economic science turning two somersaults in logic to reach an *a priori* conclusion and then calling it proof! In trying to prove too much, Mr. Atkinson holds himself up as the acme of inconsequential ridiculousness.

Even if farm mortgage debt had not grown so rapidly as assets on an average the country over, that fact would prove nothing. It does not help the people who fail to know that others succeed. While the average in figures may show prosperity on the whole, as is shown in the nation by the increase of national wealth, the great majority may be becoming poorer and poorer. There may be much suffering where wealth as a whole is increasing. No one tries to deny that as a nation we are growing somewhat richer. The trouble is that enormous sums to the credit of the few, balance the small but fatal losses of the many.

And this is the fact so far as farmers are concerned. Let statisticians say what they may, there is distress in farming communities to-day. The fact can be verified by any one who will take the trouble to travel through the rural districts of any portion of the country, and observe and listen. Men do not grumble and make themselves miserable for nothing. Unpainted, dilapidated barns, tumble-down houses, and fences out of repair; rags, lack of cash, dull, disheartened faces tell more distress than figures ever. can, especially when manipulated to deceive.

Atkinson reasons that because a certain number of present farmers came to this country a number of years ago with a certain sum of money and have now a greater sum, they made that money by borrowing on mortgage security, and farming with the capital borrowed. What a ridiculous string of inconsequence! To give any color of plausibility to his contention he would have to show: (1), that they lived on farms ever since they came to this country; (2), that all the money which they made was made by working their farms and not by going to pine woods, to cities or mines, while crops were growing or after they were gathered, as so many poor farmers in Michigan, Minnesota and Wisconsin were obliged to do; (3), that the mortgages which they are now charged with were the bases of their start in farming; (4), that it was by the use of borrowed money, and not in spite of it, that they attained their measure of prosperity. But Atkinson does not trouble about such trifles as logical sequence in his arguments on "economics." He has gone beyond that, he is an "authority."

Figures will show that the farm mortgage indebtedness is accountable for depression and failure in the farming business. According to the statements of Carroll D. Wright in a recent census bulletin, the debt on acres and untaxed mines is equal to twelve and two-thirds per cent of their value. (What untaxed mines have to do with the calculation he does not make clear.) This would make the value of acres and improvements over seventeen billions, a valuation which is not borne out by the same gentleman's figures on farm property. If our worthy statistician wanted to give something valuable why did he jumble mines and farms together and separate farms and lots? We must then take his figures of amount of mortgage indebtedness on acre property and compare it with the value of farm real estate. This value is given by Carroll D. Wright as thirteen and one-quarter billions in 1890. The debt upon it (all figures in round numbers) is two billions two hundred nine millions of dollars. On this basis the mortgage debt on farms would be seventeen and one-half per cent. The total gross product of the farms of the United States in 1890, as given by the same authority, was two billions four hundred and sixty millions of dollars. The net rate of interest on this mortgage debt was seven and one-half per cent, and the gross rate, including expenses of loan, was not less than nine per cent per annum. This would make the interest charge on the mortgage indebtedness on farm real estate alone more than one hundred ninety-eight millions per annum, or more than eight per cent of the gross product of agricultural pursuits for one year. And this does not take account of the fact that some farmers pay no interest charges, so that those who do pay, must pay more than eight per cent.

That is to say that the agricultural industries which employ, according to Mr. Wright, a capital of upwards of sixteen billions, including land values, and get a gross yearly return of two billions four hundred and sixty thousand dollars or fifteen and three-eighths per cent of the capital invested, are required to pay a fixed yearly charge of eight per cent of the gross product, while manufacturing industries with a capital of six and one-half billions and a yearly product of nine and one-half billions, cannot pay interest, rent and capital profits to the extent of ten per cent. Yet persons engaged in agricultural pursuits are comparatively prosperous! This is the stuff which Atkinson gives to an intelligent community, and which our compiler of "facts" for Uncle Sam, takes up and retails. That the business returning one-ninth of the gross product per annum as compared with capital invested, can afford a fixed charge equal to the margin of profits of the business nine times as profitable. Indeed, if we take Mr. Atkinson's figures on the margin of profits, even manufacturers with their relatively large return, could not begin to pay the fixed charges which the agriculturists of the country are called upon to pay. And we saw above that in farming, the margin of profits is really a negative quantity, without a cent of rent or interest to contend with.

But Mr. Wright goes further. He says that mortgage debts are but one-fourth of the value which they might attain without affecting the rate of interest, and there-fore, he infers, the prosperity of the farming community.

Let us see. If the mortgage charge on acre property were increased four fold, even at a gross rate of eight per cent, which would be really much below the actual, the yearly interest charge upon agricultural real prop-erty would be over seven hundred millions of dollars or twenty-nine per cent of the gross yearly product of ag-riculture. It is needless to say that that business was never undertaken which could meet such fixed charges, let alone the business of agriculture, which by Mr. Wright's own figures is the most poorly paying business in the United States. The only considerable business which approaches it in disparity between the capital invested and the yearly return, is railroading, and one-third of the railroads are in the hands of receivers. Then the railroad capital represents so much water that it is not a fair comparison. Again, the farm-mortgage debt of 1889 was seventy per cent more than that of 1880, while in that decade farm product increased but eleven per cent.

I am convinced from observation in half a dozen typical states, and inquiry in others, that the figures on farm real estate and improvements, as well as on in-crease in value for the last decade, are much too high. If farm land has increased in value since 1886, farmers have failed to find it out. The increase in value from 1880 to 1886, except in land settled in the last decade, has been largely counteracted within the last half dozen years Much land has been abandoned and much will sell for less than it would bring ten years ago. But we will accept the "facts" of our "econ-omists" and allow them to aid us in overthrowing their ridiculous conclusions. The increase in the value of land settled since 1880 is irrelevant in showing the busi-ness standing of farmers in 1890, as compared with 1880.

Whether there is a margin between the value of the property mortgaged and the amount of the mortgage has nothing to do directly with whether the business can bear the burden and still be profitable. It is the relation of the interest charge to the annual product, or what is left of it after paying necessary expenses of production, which must decide as to whether this or that business can bear its load of debt. To be sure, the comparative size of the debt is now an indication of what the interest charge is, but were no interest charged the whole capital might be borrowed without affecting the prosperity of the business. While the lender may safely base his estimates on the value of the security, the borrower must base his on the margin of profits.

As to the question whether the agricultural product is sufficient to bear the present interest charge and still leave enough to pay the necessary expenses of production, there can be but one conclusion: that it is not. The figures and common sense of the matter agree with observed facts. Agriculture is overburdened and unprofitable. There is no miracle by which the least paying business is most prosperous in a period of general distress.*

But we will take the mortgage debts as a whole. While the prosperity of the farmer is important, it is no more so than that of any other honest toiler. Taken as a whole, we see that while our assets have increased less than fifty per cent in nine years, our fixed debts have increased more than one hundred and fifty-six per cent. Any one with ordinary common sense, not to speak of economic knowledge, who can find a sign of prosperity in this, is more worthy of congratulation on his optimism than on his intellectual powers.

But while this ground was being lost have we not heard prosperity cried from the housetops? Certainly, the capitalist was prosperous, he was increasing his income at an unheard-of rate and his voice is loudest.

---

*Consider for a moment that according to census figures the 43½ millions who live outside of cities subsist on a gross yearly product of 2,460 million dollars, while the 18 millions who live in cities have a gross product of about six billions of dollars per year, besides the profits of merchants and the remuneration for professional and personal services, and then ask are farmers prosperous? Three times too many people are trying to live on farms.

It was the interest charge and the rent tribute that was piling up the big aggregate on the wrong side of the ledger. And judging from the relation of the increase in indebtedness to the increase of national wealth, all of the interest charge did not go to American capitalists, either. Nothing except the principle of interest-taking can account for a decline in the fortunes of the producing masses, in times of profound peace, vigorous effort and abundant rewards. The wealth was certainly produced; enough of it to make the whole people prosperous. The increasing mortgage indebtedness must then be an index of its distribution between the classes. The rich have certainly grown richer. Then it is the poor who must have suffered, and they suffered through unearned incomes of the rich. For nothing *earned* by any one can make another poorer.

We have thus seen that the taking of rents and interest is amply capable of explaining all of the considerable disturbances in the industrial and financial world. On the same basis we might explain the growth of the tenant farmer, the tenant population and slums of our cities, and a thousand lesser troubles with which we are menaced. We can easily explain on this basis the fact that luxury and squalor go hand in hand. None of these phenomena can be satisfactorily explained on any other theory. Our industrial, including our financial system, is based on the taking of interest and rents, and if these principles are right, and are applied, our system should be faultless. The principles are certainly applied, but the system is certainly not faultless. Let the plutocratic apologist afford an adequate explanation. It has not yet been done. The nearest approach to it is the generality that these troubles are an incident of our civilization. A meaningless platitude, unless we show how they are. This I have endeavored to do. I have at the same time shown that my theory bears the crucial test of truth; i. e., agreement with observed phenomena. It is in no place out of harmony with observed facts.

CHAPTER XXIII.

"Truth crushed to earth will rise again."

PERSONS may pronounce it strange that the world has waited until this day and generation to discover the evil of interest-taking. The fact is, it has not. Many important discoveries have been put off until the nineteenth century, but this is not one. Although the evil was detected it was never explained, and hence people foolishly came to the conclusion that it was imaginary.

From the earliest dawn of history, we find the most advanced thinkers and law-givers combating usury, which term until recently was used to designate the taking of all increase. It was fought in every civilized state. It was the ruin of more nations than all other causes put together.

By the Mosaic law, usury was strictly and unequivocally forbidden. Deut. XXIII. 19: "Thou shalt not lend upon usury to thy brother, usury of victuals, usury of money, usury of anything that is lent upon usury." This is sweeping enough to meet any case. But in the next verse the law-giver goes on: "To a stranger thou mayest lend upon usury, but to thy brother thou shalt not lend upon usury, that the Lord thy God may bless thee in all that thou settest thy hand to in the land whither thou goest to possess it."

The latter passage is taken by some to break the force of the interdict of the former, but it will bear no such interpretation. The believing Jew has always considered his own people his brothers and all others as different beings, having no rights which a Jew is called upon

140

to respect. All beside the chosen race are delivered into the hands of the elect for their special benefit, and are therefore fit subjects for all sorts of extortion. The modifying clause is but an index of early Jewish character. If it were otherwise it could not belong to a people among themselves the most just on earth, to others the most unscrupulous and vindictive.

And it was not the law-givers alone who looked askance on usury. The populace were in full sympathy with these laws. We read in Nehemiah V. 7: "Then I consulted with myself, and I rebuked the nobles and the rulers and said unto them, Ye exact usury, every one of his brother, and I set a great assembly against them." And further on he continues: "I likewise and my brethren and servants might exact of them money and corn; I pray you let us leave off this usury." Still further on he says: "Restore the hundredth part of the money, the oil and wine which you exact of them."

The last passage effectually disposes of those who hold that the early Jewish law permitted the taking of some increase but condemned excessive usury. One per cent is as small an increase as we could consider at this age of the world, yet the prophet condemned it. He was an authoritative interpreter of the law.

The Jews placed the usurer in the catalogue of the hopelessly incorrigible, and in Psalms XV. 5 it enumerates among those who may be appealed to: "He that putteth not out his money to usury nor taketh reward against the innocent; he that doeth these things never shall be moved."

Proverbs says: "He that by usury and unjust gain increaseth his substance, he shall gather it for them that shall pity the poor." Even the giving of usury is condemned by the prophets, as Isaiah says: "The Lord maketh the earth empty, as with the taker of usury, so with the giver of usury to him."

Jeremiah looked upon usury as a crime worthy of divine visitation, for in his lamentations he says: "Woe is me, my mother, that thou hast borne me a man of strife and a man of contention to the whole earth; I have neither lent on usury nor men have lent to me on usury; yet every one of them doth curse me."

Ezekiel numbers among the just who shall live: "He that hath not given forth upon usury, neither hath taken any increase."

The unjust son is he "that hath given forth upon usury and hath taken increase." The just son is he that hath not received usury nor increase.

The wicked city is thus addressed by the prophet: "Thou hast taken usury and increase, and thou hast greedily gained of thy neighbors by extortion."

We find running through the whole Bible story of the Jews a thread of policy against usury, yet we find that in the days of Nehemiah, the taking of usury on mortgages made in years of short crops had reduced a portion of the population to slavery, and that he appealed to the people and abolished usury, even to the taking of a hundredth part. The history of the Jews was like that of other nations, but they were wise enough themselves to learn the lessons of history and to abolish the principle, the most potent in bringing about internal disorders in the state. The constant reference to usury in the Old Testament shows it to have been a question of the utmost importance among the Jews.

On the other hand they well understood the power of usury in appropriating the substance of other peoples, and in all ages have by this means controlled the finances of the world. The breed of barren metal acting through the Rothschilds of to-day is the governing power of the civilized world.

Greece early felt the power of usury. Having experienced the impossibility of having all their demands satisfied by the productive activity of the laborers of ancient times, and not satisfied with the substantial benefits of slavery without its responsibility, the usurers of ancient Greece contracted that the persons of their debtors should stand as security for debts. In that way we find that at the time of Solon the former independent proprietors of Athens had lost all of their property to the rich and were, to a great extent, enslaved through the taking of usury. Interest-taking had given all the power in the state to a small plutocracy. Solon, or the laws attributed to him, cut the

Gordian knot by canceling all debts founded on the security of land or the person of the debtor, and providing against arrearages.

The history of Sparta is similar to that of Athens.

While combating the effects of usury, the Greeks did not seem to understand clearly its ethical status. Aristotle accounted for the wrong of usury by the barrenness of money. Had he gone a step further and explained it by the barrenness of all wealth, he would have stated a truth that never could have been successfully controverted. Those who contend with him try to show that other sorts of wealth are not barren, but their contention is so thoroughly fallacious, that it is a wonder that it ever passed current with disinterested thinkers. A flock of sheep left to the mercies of their natural enemies would not be wealth at all. Neither would they increase to any extent. It is the labor of their care which keeps them immediately useful for satisfying the wants of man, and therefore wealth. In a wild state it would be the labor of hunting them which would make them available. There is no increase, no wealth without labor.

Rome had an experience similar to that of Greece. The sturdy yeomen of earlier Rome became, through usury, hopelessly in debt to their richer neighbors. The legislation of the twelve tables was intended to meet this evil, but failed, and the free Roman farmer was destroyed. Many of them were reduced to slavery. So pitiless were the usurers of Rome that the war tax of Senna increased six fold in fourteen years. When Cæsar came into power, he practically adopted the legislation of Solon, but whether not enforced or inadequate, it finally failed of success. While in the city of Rome interest was but four per cent, in the provinces it rose as high as twenty-five or thirty per cent. Justinian fixed the rate of interest at six per cent and made arrears non-collectible, but after a time, we find the usurers ruling and the working people of the nation falling into decay. This was the first, as well as the last, step in the downfall of the Roman empire. Rome is an example of power without conscience, carrying within it the

seeds of its own destruction. A great nation can no
more be maintained on wrong than a temple can be
built on sand. Usury is the greatest industrial wrong
ever perpetrated.

Usury was always a crying evil in Rome. Many of
the great leaders spoke most vigorously against it. It
was said to be the cause of the greater portion of the
sedition and discord which perturbed the state. Cicero
says that when Cato was asked his opinion of usury,
he made a Yankee-like response by asking the question-
er's opinion of murder. Seneca and Plutarch are both
recorded as inveighing against it, and Tacitus notices
its mischief to the state. To the time when usury lorded
among thousands, Rome dates its decline.

At times in Rome the taking of usury was treated as
an aggravated species of theft and punished accordingly.
There are instances of the death penalty having been
inflicted for the taking of usury. This ridiculously
severe punishment, to be sure, served no useful purpose,
but rather served to bring about a revulsion in favor of
the persons against whom it was directed.

There is but one moral from Roman financial history:
The Romans failed to destroy the taking of usury and
usury-taking destroyed Rome.

There are several reasons why the ancient laws
against usury were not effective. The money of the
ancients was founded on a metallic standard entirely,
and with such a standard it is impossible to prevent
interest-taking completely. Judges became venal and
failed to enforce the law or connived at its violation.
The persons most influential in government were most
benefited by the taking of usury, and the people who
were filched from were too ignorant to understand the
situation or an adequate remedy for the trouble. The
latter principle is found to be at the bottom of the per-
petuation of all abuses which have cursed the nations
of the earth. But more of this hereafter.

Without going into the history of usury in detail, we
may say that it has been fought continually down to
the present time. The community of interest of the
early Christians strictly forbids it. Of the fathers of

the church, we find St. Ambrose, St. Augustine and Leo the Great vigorously preaching against it. St. Bernard made an enthusiastic crusade against it. Pope Alexander III. joined in condemning it. Luther roundly and characteristically scored the usurer. Melancthon, Beza, Musculus, were all counted among its enemies. With few exceptions, churchmen opposed it down to the sixteenth century, and for a much longer period many of the leaders in church and state were unalterably opposed to it.

So bitter were the English populace against the usurer, that from the time of Alfred down to the thirteenth century, detected usurers were often mobbed and roughly handled. The statute of Merton, passed 1235, was the first formal English law against it, although it was an indictable offense at the common law. In the latter portion of that century, the Jews were expelled from England for their usury-taking propensities. Indeed, this trait of the Jews accounted for much of the harsh treatment which they received at the hands of mediæval Christians. Shakespeare powerfully sets forth this truth in the "Merchant of Venice." It is an excellent mirror of the times, and Antonio, "who neither lends nor borrows by giving nor by taking of excess," is a type of the honorable merchant prince of the Middle Ages.

Through all of the writings of the immortal bard we find usury held up to public scorn. In "Lear" the usurer hanging the cozener, is cited as one of the grim ironies of civilization. This is reason and impertinency mixed. We find the usurers spoken of in "Timon of Athens" as "bawds between gold and want." In the "Winter's Tale" we find the usurer spoken of in the most contemptible manner. Indeed the works of Shakespeare bristle with references to the wrongs of usury. Bacon voices the same ideas.

As the Jews were hated for usury-taking, so they hated the Christians, for trying to ruin their trade in breed of barren metal. It was because Antonio lent out money gratis "and brings down the rate of usance here in Venice" that Shylock wished to catch him once upon the hip. Shylock and Antonio are types.

So strong was the sentiment against the taker of increase that for a time he was denied consecrated ground for burial. Dr. Wilson, a writer of this time, rather naively but savagely writes: "For my part I will wish some penall lawe of death to be made against these usurers, as well as against theeves and murtherers, for they deserve death much more than these men doe; for these usurers destroi and devour, not onlie whole families, but also whole countries and bring folke to beggerie that have to doe with them." .

The laws did not go so far as that. There were severe penalties, however, attached to the taking of usury, and they served as a more or less effective restricting influence. The first penal statute against usurers was that of Edward III. It provided for forfeitures and restrictions. It was followed by still more rigorous laws, and still others were passed in the reign of Henry VII. A penalty of 100 pounds was attached by the latter to the crime of usury, as well as forfeiture of the principal and a depriving of the usurious lender of the privilege of again engaging in the business.

The statute of Henry VIII. was the first to treat usury as the taking of *excessive* interest, and by this law lenders were allowed to collect not more than ten per cent for their money. Statutes in the same line were passed up to the reign of Anne and interest thus gradually reduced to five per cent. The statute of Anne was an especially strict law and seems to have been carefully administered with a view of enforcing its provisions, and to have been very effective in the suppression of the taking of interest beyond five per cent.

The laws passed since the time of Anne have been in the direction of relaxing the strictness of the earlier statute of usury, and in England and all countries using English law as a basis, the laws against excessive interest-taking have been greatly modified in that direction. The statute was modified in England, not because it was ineffective, but because it was rather too effective and did not agree with the theories of the leaders of English thought.

France had an experience similar to England, and

the policy of that country was first in the direction of suppressing usury altogether, and then toward allowing it to a limited extent, under the term of interest. Here, as in England, the effectiveness of the law was determined by the intelligence displayed in its drafting, and the zeal of judges in enforcing the law. It is a notorious fact that everywhere the usurers are the persons who most loudly declaim against the effectiveness of laws against usury.

This very general sketch serves to show that usury has always been questioned by some of the best minds of the world, and it was only in a civilization where material prosperity of a few was more important than justice to the many, that usury was finally sanctioned by law. Laws are usually an embodiment of the interests, supposed or real, of the persons in control of government, modified by a few concessions to the multitude. The concessions to the multitude in several of the above cases were the legal straws against usury.

The fact that usury was banned by law under a civilization seemingly lower, and recognized by law under a civilization seemingly higher, proves nothing as to the ethical right or wrong of the custom. There are numerous laws founded on wrong which are now and have ever been tolerated. We find slavery a recognized institution in all the nations of the world. It was supported and upheld by law, down to the middle of the nineteenth century, in countries most advanced in civilization, yet no one would have the hardihood to say that slavery is right. It may be introduced again, yet that will not prove it right. The slavery brought about by interest-taking is just as worthy of condemnation. The most audacious will not for a moment assert that war is right in itself. Its utility to the world could not be established by a library of argument, yet the laws of all nations recognize war, and war is the *raison d'etre* of many governments. Man prides himself on being civilized, a being of superior intellect and sentiments, and shudders at tales of cruelty in beasts; but neither the jackal, the lion, the tiger or the wolf have ever shown such utter savagery and unreasoning cruelty

as the glorious soldier of the civilized, Christianized
nations of to-day. Their law-sanctioned action would
be considered'outrageous, judged by the standards of
the most blood-thirsty beasts. One might fill volumes
with a catalogue of wrongs which obtained the persistent sanction of law. Every generation has the right to
examine all that has been handed down to it and to
verify for itself the conclusions of those gone before. It
is not only a right but a duty, for on this depends all
progress. On ethical progress must depend all lasting
material progress. Overturned and demolished idols
are the milestones on the highway of ethical progress.
No legal sanction can sanctify a wrong and no amount
of popular or political combination can make false the
true and right. If all the philosophers and mathe-
maticians who have ever lived should come in long procession and declare that half of an object was greater
than the whole, I should say, "Avaunt, idiots, you speak
falsely!" The truth which, I have been affirming rests
on as secure a basis, and we can afford to fly in the face
of convention in asserting it. But as I have shown, the
opposers of usury are not solitary. Not only have the
best laws of the best peoples been directed against it,
but a long line of thinkers from Aristotle to Proudhon
have sought to open the the eyes of the people to the
wrong of interest-taking. It was only because they
failed to understand fully the absurdity implied by the
foundation principle of interest, that they failed of
their purpose.

CHAPTER XXIV.

"To him who has shall it be given."

In the light of ethics, history and economics, inter-est-taking is indefensible. It has from the earliest his-tory been the bane of civilization, it still remains the leaven of discord and destruction. The problem is how to destroy interest-taking. In determining what makes interest-taking possible, we may arrive at a method by which the practice may be destroyed.

The necessity of borrowing wealth for use in industry, gives to those persons who have accumulated surplus the opportunity to exact interest. When interest is once exacted, it, by its principle of geometric increase, keeps the lender in position to exact continued interest and the borrower under the necessity of contracting continued loans. Borrowing is made necessary by the progress of civilization in the present lines; by the present financial system, by the present basis of money.

Millions are born into the world every year. They have wants which must be provided for. The gratuitous gifts of nature are not sufficient to support in ease the teeming millions of the earth. Population has increased to such an extent that its wants, born of civilization and the necessities of the climate into which its increas-ing numbers have driven the human family, cannot be supplied even by the efforts of unaided man directed to the free gifts of nature. The best methods, the best machinery must be employed by every individual if he would succeed among the increasing multitude, in the ever fiercer struggle for existence.

149

Tools and machinery are as necessary to modern industry as is land. Without them, we should not have modern industry but the crude methods of the savage. The laborer with the spade and reaping-hook cannot compete with the laborer with the gang-plow and binder. The power-press daily has driven the hand-press sheet to the wall. The power-loom and the spinning machine have driven out the distaff and the hand-loom. The cotton-gin replaces the hand-picker, the thresher the flail, the roller-mill the mortar, the steamboat the oar, the telegraph the courier. If one have not these tools he must borrow them or wealth to trade for them, or he must work for another, just as surely as he must work for another if he have not access to land. It is only by intelligent labor with the best appliances that mother earth can be prevailed upon to satisfy the needs of the millions who swarm upon her breast.

The majority of those born to the earth have not the implements of toil and are therefore constrained to borrow them, or, what is the same thing, work for another and give him the profits of their toil, retaining for themselves bare subsistence only. This necessity for borrowing must always exist to a greater or less extent. Under the present organization of industry, the great majority of those who toil use implements not their own, implements on which they are obliged to pay interest or capital profit. They are almost absolutely at the mercy of those from whom they borrow. Subsistence is a necessity which cannot be turned aside. Millions feel that necessity. It is then not difficult for the holder of wealth who wishes to get the most out of his possession to impose terms on the borrower, just falling short of pressing him into destitution. A man with starvation staring him in the face cannot haggle about wages, especially when there are hundreds of others at his side ready to work at the figures offered. The man in need of money to pursue his business, money which he must have or fail immediately, is not in position to drive a bargain with the money-lender. He must and does come to the terms of the latter.

But in the ordinary course of business, the necessity

for borrowing does not depend immediately on the necessity of engaging in remunerative toil. The financial system of this country and all others renders borrowing by business men absolutely necessary to carry on their occupations.

They must do it in order to gain control of the mechanism of exchange, the vehicle of all intercourse between man and man. Under the laws of interest and rent the capitalists of the country, as such, each year receive an amount of wealth so large that they are able to save from it a sum greater than the yearly net increase of the wealth of the nation. It is theirs, they may do with it as they see fit. If capitalists conclude to hoard their wealth, that which they receive each year is doubly sufficient to command all of the circulating medium of the country. Interest and capital profits are contracted for in money; money is the only form in which wealth can be hoarded without loss. It is money, then, that the capitalist will naturally demand. From his position at the head of the principal financial institutions of the country and his intimate relations with all others, he is in position to command the money which he wishes. The capitalist is then placed in control of the currency of the country. He may withdraw it from circulation, and the only way in which it can be restored is by borrowing from him at substantially his own terms. But money represents wealth, and by taking the money from circulation the capitalist ties up in the hands of producers wealth equal, at least, to the face value of the money which he controls and hoards. In practice, he actually ties up several times that amount of wealth, for a given amount of money is usually capable of supplying a medium for the transaction of exchanges to many times its face value.

That such is the effect of hoarding money, no one knowing the elements of economic science will attempt to dispute. It is not necessary to go into an argument here to establish its truth. J. S. Mill has conclusively proved that, under normal conditions, demand and supply are equal and that production is the cause of demand. Demand, says Mr. Mill, implies the wish for an article and the ability to purchase it. The production of the equivalent of the article desired or the taking of that equivalent from him who has produced it, is the only means by which the second condition of effective demand can be fulfilled. Evidently the persons who pay the billions collected annually for rent, interest and capital profits, deprive themselves by just that amount of the ability to purchase, and there is just that much less effective demand for articles used by the productive toilers. Then the amount which capitalists decide to hoard is represented by goods on the market for which there is no demand. This wealth which the hoarded money called for, is being cared for by others who are obliged to bear its deterioration and the risk of holding, and the capitalist can afford to wait. He is master of the situation. He has the ability to purchase but not the desire, and other people who have the desire to demand the goods on the market, but have deprived themselves of the ability to do so by the payment of unearned incomes to the capitalist, must secure that right by borrowing back at the terms of the capitalist the ability to purchase. If they do not do this they must look upon glutted markets, languishing trade, suspended industries, idle laborers and starving multitudes. These are the indications that the active toilers of the country can no longer meet the conditions imposed upon them by the lending capitalist, or that the latter prefers to hold on to his unearned gains until the industries deranged by his inordinate demands are placed on a normal footing, and there is less risk of loss. He waits until the panic has passed its acute stage, and then, like the tardy doctor, offers his medicine when the patient has, of his own resources, recovered his normal conditions or is beyond aid. If interest and rent were not collectible by private parties, the fund which they represent would be in the hands of the active toiler, supply and demand would always remain equal, there would be no stagnation, "glut," or "overproduction," laborers would be steadily employed.

When a toiler produces an article, he does so because

he wishes to consume that article or to trade it for something to consume. The toiler with whom he trades, if he trade directly, produces what he does with the express purpose of supplying the demand of the first toiling producer. By the act of producing each creates a demand for what the other produces. The demand is, in each case, just equal to the amount produced. But this is all upset by the non-producer coming in and taking a share of the production. He deprives the producer to that extent of effective demand and makes the goods which his fellow-producer creates for him a drug on the market. The interest-taker has the effective demand which formerly belonged to the toiler, but he is usually obliged to use but a portion of it, and the rest allows the goods intended to supply it to become a drug on the hands of their producers. Then the wants of the interest-taker are not the same as those of the toiler, and other sets of toilers are put to work on other industries to supply the wants of the taker of unearned incomes. So far as the great body of productive workers are concerned, the goods made to supply the takers of unearned incomes are a total loss.

If Jack manufactures shoes which he wishes to trade for William's flour and Henry's meat, it is to the interest of William and Henry, as well as Jack, that he retain all of the shoes which he manufactures to trade for their flour and meat. For, if a portion of the shoes are taken from Jack in unearned charges, that leaves him able to buy just that much less flour and meat. The taker of these unearned charges may or may not buy flour and meat to the extent which Jack would have done had he retained his whole product, but as a matter of fact, he does not. Then when we consider that a like amount is taken from Henry and William, we see that the production of each gives him effective demand, diminished by just the amount taken from him in unearned incomes, and the toilers as a body are worse off by just the aggregate amount taken from them in unearned incomes.

The law of supply and demand will work only where every one retains his own. In an economic organization where a few are allowed to take a portion of the substance of the many without giving value received, the law is modified and its effect appears distorted. The law of organisms is to develop to maturity, yet ignorance or casualty makes many individual organisms abortive, and kills many others in an undeveloped state. Just so the wrongs of our economic organism so modify the natural laws governing it that the law that supply is equal to demand and that production is the cause of demand, is distorted, demand becoming the cause of production and the supply and demand being so poorly balanced that we have glutted markets in the midst of starvation, and idleness on the threshold of poverty. The business of the statesman of to day is to remove the artificial modifications of natural law and allow our suffering industries to reassume their natural equilibrium.

CHAPTER XXV.

The hand that holds the dollar is the power that rules the world.

THE practical manifestations of financial panic are very materially modified by our currency system, which gives power to the rent and interest-takers to control our medium of exchange, and thus tie up business more effectively than could be done under a rational currency system.

When the currency of the country is monopolized, the laws of supply and demand are no longer operative; markets seem to glut while a portion of the inhabitants of the country are on the verge of starvation. Production seems to cease or drag along, just at the time when there is most need of wealth of various sorts for the support of the race. It is possible for the financial classes to monopolize the currency, because of the fact that they collect, in a few months, a sufficient amount in rent and interest to equal the whole volume of circulating medium. The basis of the monopoly is interest-taking, but the monopolization is made easy by the adaptability of the currency system to that very end.

The hoarding of the capitalist at critical times has an even broader effect. It deprives the industrial world of the instrument of exchange called money. If a class of men had control of all of the scales in the country

155

and refused to allow them to be used, the result would be a serious embarrassment to business. But money is a mechanism much more necessary to our exchanges than are scales. It enters into every transaction, both as a measure of value and as a means of transferring property. While the hoarding of scales would embarrass that class of exchanges only where the goods are sold by weight, the hoarding of money would embarrass all exchanges and virtually paralyze business. Then, were scales cornered, we might soon produce new ones of the material at hand, but money is practically a fixed quantity, probably growing smaller in proportion to the service required of it. So close and exclusive is the private government-sanctioned monopoly in money, that if the meager supply of gold can be cornered, the law decrees that the hold-up is complete, that nothing else shall take its place, and that the hapless industrial victims must submit to the demands of the financial road agents.

This is not an imagined danger, but is seen to be very tangible and persistent at times of financial distress. Everything but money seems plentiful enough, but that is so scarce that business refuses to move.

The explanation is not difficult. We have become accustomed to make our exchanges with money and we cannot do without it. It is necessary, both as a denominator of value and a medium of exchange. Suppose, for an instant, that A had a wagon, B a horse, C a pair of oxen and D several sheep. A wishes to sell his wagon and get a horse, B wants to sell his horse and get a flock of sheep, C wants a wagon for his oxen and D wants a yoke of oxen for his sheep. The demands of each might be supplied within the circle, but as no two could trade directly there must be some medium to carry on the exchanges, else great difficulty and annoyance would result. Money, then, has a power as a mechanism outside of and distinct from its power as a representative of wealth, and may have a paralyzing effect on industry far beyond the actual demand represented by the bulk hoarded.

Viewed from whatever standpoint, the taker of rent

and interest or capital profit has the key to the situation. The financial group controls the industries of the nation. If industries wish to proceed they must borrow the mechanism of exchange from the landlord and usurer, and largely on his own terms. For in the loaning of capital we have the closest trust in the world. The money combination is so far-reaching and effective as to be invulnerable. The capitalists hold the fortunes of the toiling world in their hands and can dictate terms. What these terms are in times of need has been illustrated in every financial panic ever known. In that of 1893 money was held at from ten to seventy per cent or refused altogether, unless to the gambling class who could use call loans. The laborer must take less wages, the active business man less for the wages of management or superintendence, but the usurer must have his pound of flesh.

It is needless in this connection to divide interest into interest proper, capital profits, etc. After enough has been set aside to keep up the capital engaged in business, anything that is paid for the services of capital and not for the actual personal services of its owner is an unwarranted charge on industry. It rests on the false principle underlying interest: that capital has the power of production and should be remunerated therefor. We have seen the truth of the proposition laid down by Adam Smith, but never followed to its logical conclusions: That the laborer produces all wealth. The logical conclusion is that therefore all wealth is rightfully his.

We must, then, banish all taking of increase for money lent, and substitute for the present metallic currency a money which cannot be cornered. For admit the right of capital profits, and those who control the surplus wealth of the nation can, by contracts wrung from the necessities of their fellow-men, keep control of all possible net earnings of the nation's industries and therefore all the currency of the country. They may lend and re-lend, but on such conditions that when the day of reckoning comes they are in the same advantageous position as before

It is needless to catalogue instances of the conditions to which attention has been called. They are manifested over and over again in every period of distress. The people have become so accustomed to money stringency, glut of markets, over-supply, suspension of industries, timidity of capital, that they are accustomed to look upon these phenomena as the causes instead of the symptoms of financial disorder.

Grave financiers voice the same error, seemingly unmindful that there must be a common cause for all these indications of disease in the body industrial.

Our present money seems especially destined to be used as the tool of the usurer. Its adaptability for cornering and contraction of volume is vastly superior to its adaptability for any other purpose whatever. A glance at the facts is sufficient to substantiate these statements.

Political economists hold that the value of the dollar is inversely proportional to the number of dollars in circulation, the volume of wealth being constant. While it is doubtful whether the law is mathematically exact even with the present monetary system, and while it is certain that it would not hold at all in a scientific currency system, experience has shown that with a commodity money, that statement approximates the truth. Prices measured in money fall with contraction and rise with expansion of the currency. If the law be true, the money of the country should bear a fixed relation to the wealth of the country, or at least to the volume of that country's exchanges. The volume of a country's exchanges depends on the volume of that country's wealth and we may take the latter as an approximate criterion of the volume of money required.

Placing the national wealth of the United States at seventy billions of dollars, gold basis, and the money in circulation at present at $1,740,000,000,* a recent estimate of the Treasury Department, there is in this country one dollar in currency for every forty dollars in wealth; placing the currency of the country on a metallic basis, as all clamor for; i. e., using our available stock of gold and silver at its present money value,

* The half billion idle in the United States vaults is not in circulation.

we would have but one dollar in currency for every seventy-two dollars of national wealth. Making gold the sole money of the country, as we have just attempted to do, would give us but one dollar in money for every one hundred forty-five dollars in national wealth. This estimate is made on a basis of gold stock of six hundred four millions, and a silver stock of six hundred fifteen millions, allowing one-fifth for national bank and United States Treasury reserves. If money be scarce and prices low enough to cause financial embarrassment with one dollar in currency for every forty dollars in national wealth, it would be quite out of the question to carry on business with one dollar in currency for every one hundred forty-five dollars of national wealth. Prices measured in money would fall to less than one-third of their present level, and it would require three times as many goods to pay fixed money obligation as it would have taken at the time the debt was contracted. The result would be fearful to contemplate. The financiers of the country would by one stroke get control of all wealth.

Nor would there be any prospect of supplying the deficiency in the gold currency. For the last decade our national wealth has increased on an average more than two billions per year. To maintain the present ratio of the volume of currency to the national wealth and hence the present prices and conveniences of exchange, would require a yearly addition to the circulating medium of about fifty millions of dollars, which, with the proportional reserves, would amount to sixty-two and one-half millions yearly. The entire gold product of the country is but about thirty-three and one-half millions yearly. The coinage per year has so far been about equal in amount, but there is no prospect that this will be allowed to continue. Indeed, if gold were not bolstered up to such a high price by its preference as a money metal, the entire product would be used in the arts.

We cannot draw on the rest of the world, for the rest of the world is in little better position to make gold its money. Placing the wealth of the world outside the

United States at two hundred twenty-five billions and
the circulating medium, exclusive of the gold reserve
of one-fifth, at 6,816 millions, there would be an average in wealth of thirty-three dollars for every dollar in
money. On a coin basis other nations would have an
average of one dollar in money for every forty-four dollars of wealth, and on a gold basis, one dollar in money
for every ninety-five dollars in wealth. Estimating
the average yearly increase of wealth outside the United
States at four and one-half billions of dollars, it would
require a yearly addition of one hundred fifty-three millions of dollars to foreign currency to maintain the
present relative volume of currency and hence the present prices and commercial facilities. But the entire
gold product outside of the United States was valued
at but ninety-eight millions of dollars in 1892, and
twelve millions less was coined in 1891, the last year
for which I have figures. Then if the present volume
of money is not too great, and all will agree that it is
not, and if it is wise to maintain at least as great a
relative volume, an exclusive gold or an exclusive silver
currency for the United States or for the world is quite
out of the question.

Fluctuation in the volume of the currency means
fluctuation in the money standard, with all of its consequent evils. But a fluctuation in the money standard
may occur without a fluctuation of the volume of money.
It must occur with a money standard or value denominator based on a single commodity. Gold has never
been a fixed standard. As compared with the wholesale prices at Hamburg of one hundred staple commodities (as collated by Dr. Soetbeer, the economist), gold
has appreciated in value more than thirty per cent since
1873. The "Economist's" list of twenty-two index
articles, giving average prices in London for twenty
years, shows the purchasing power of gold to have appreciated from thirty-five to forty per cent within that
period. A list of prices of standard articles in New
York would show a like result. This proves beyond a
doubt that the gold dollar has radically appreciated in
purchasing power and that its exclusive adoption as a

money standard would entail all the evils of an appreciating currency.

It is entirely irrelevant to say that gold has remained stationary and other things have fallen in price. A just money standard should adapt itself to the average fall in prices of other commodities or the financial group alone will get the benefit of improved processes of industry, and the value of all inventions will go into the pockets of the money-lenders.

Silver having been a simple commodity, but a favored one, in the United States and in many of the European countries since 1873, sympathized pretty fully in price with other commodities, until at the beginning of 1893 the concerted move of the gold forces against its use as a money metal threw a large supply of the commodity on the market and depressed the price below the normal standard. To be sure, increased production made the fall more marked. The adoption of the single silver standard would have the immediate effect of debasing the currency and driving the gold out of circulation, but would be followed by a reaction in which the price of silver would probably double and the money standard be appreciated to nearly its present mark. A single silver standard would mean wide fluctuation and consequent widespread ruin.

There is no possibility of making a fixed quantity of any commodity a non-fluctuating money standard. The price of a commodity which can be produced in any desired quantities ultimately depends on the cost of production, but the supply of a commodity limited in quantity, and the fixed stock of which is relatively large as compared with the yearly product, depends almost exclusively on supply and demand. The demand for money may at any time reach the volume of all of the exchangeable wealth in the country, while at other times it may be much less. No one commodity in a country can equal all other commodities, and hence the demand for a commodity which is needed to represent all other commodities must exceed the supply. In this case the price must go up. This is what is actually happening every day. We have tried to make six hun-

dred millions in gold supply the demand for two thousand millions in currency. Hence gold is sought after,
is scarce and dear. We could make lead dear in the
same manner, or silver or any other commodity, but
not so rapidly, as the supply is not so meager.

A recent writer has argued that the volume of a commodity currency has nothing to do with its price. If
its price depends on supply and demand, then the demand is greater, he argues, the greater the volume of
the currency, and the price of a fixed quantity should
be greater, and *vice versa*. It is not the absolute demand, or the demand for metal to coin which alone
fixes the price of that metal, and therefore the value of
the money standard; but the volume of the metal coined
as compared with the volume of the metal available for
coinage, or the total volume of the metal or commodity
as compared with the total demand for the commodity
for coinage and other purposes, which fixes the price
of a fixed quantity of the commodity, and hence, the
money standard. Now the smaller the volume coined
at a certain price, the smaller the volume available for
coinage at that price, while the smaller the amount of
the money commodity coined as compared with the
volume of wealth which it represents, the greater demand for the coin and consequently the greater the
demand for the commodity to coin and the greater the
value of the unit. The relation of demand to supply
fixes the price of the unit. Gold and silver are subject
to all fluctuations of other commodities and hence
neither can ever be a fixed money standard. As a matter of fact gold and silver do respond to the law that
the value of the dollar is inversely proportional to the
volume of the circulating medium, at least above or
below a certain fixed volume.

It would be the same whatever commodity money we
had. If lead were made the sole money of the country,
the demand for lead for coinage purposes would greatly
affect the price of that commodity.

The law of supply and demand, as I have stated
above, holds for money as well as other articles. In
this, too, under normal conditions, supply and demand

must be equal. Now, if with a certain volume of money and a certain volume of wealth the conditions are normal, if supply and demand are equal, then with any other volume of money with the same volume of wealth, supply and demand must be unequal; or the only way to keep up the equilibrium is to make the money unit more or less valuable. This is what actually happens. If Jack holds one hundred dollars and it represents a certain fixed quantity of wealth, each dollar will buy one-hundredth part of that wealth. Let his dollars be increased to two hundred, while the wealth which they represent, or which can be purchased for them, remains the same, and each dollar held by Jack will purchase but one two-hundredth of the wealth. In other words, the money unit has lost one-half of its value. There is no contradiction in the statement that the value of the money unit is regulated by supply and demand and also by the comparative volume of currency.

The problem to be solved in establishing a scientific currency system, is to devise a money, the supply of which will always just equal the demand, and the value of a unit of which will therefore always remain constant. No metallic currency can satisfy these requirements.

Indeed, if our currency were especially invented for rapid and certain fluctuation, both as to volume and the value of the money unit, it could not be better fashioned than now. With a currency of more than two thousand millions of dollars based entirely on considerably less than one-third that amount of gold, the system is more mercurial than mercury itself. The paltry stock of the yellow metal is supposed to supply United States Treasury and bank reserves and to redeem the remainder of the currency on presentation; yet it is but a fraction of the value which it is supposed to pay. How wonderfully elastic that hoard of gold! Financiers explain the necromancy by which gold can support and redeem a currency of three or four times its volume, by saying that it is not intended that the currency is to be presented for redemption. That is to say: "If you don't want our gold you may have any amount of it, but if you want it, under no circumstances can we let it go. We will suspend." The fact is that the explanation is no explanation at all. The real explanation is that two-thirds to three-fourths of the currency of the United States is irredeemable paper and silver, with the gold unit as the money standard.

This sort of currency is allowed to remain because it serves the purpose of the usurer. If it happens to be to his interest, as it usually is, to make the dollar more valuable, he can just demand gold for a portion of the currency supposed to be founded on it; others will follow suit. The price of the scarce commodity goes flying upwards, and down go all other prices, while the volume of the currency goes on shrinking. The usurer's money will then lend for more, his fixed credits will be worth more. He reaps the harvest. So does the speculator. It is not necessary to remind the student of financial history of instances of this sort of manipulation.

Of course the demonetization of silver made the volume of this unsupported currency greater and was in that way an advantage to the manipulators of money.

Bimetallism will not cure the trouble. It is merely a sop to those who do not implicitly believe in gold as the divinely constituted money of the world. While gold and silver are used as money they will be superior to all other sorts of property both for hoarding and lending, and for that reason if for no other, will always in time of pressure be collected and hoarded by those who control the surplus products of the nation and dominate its financial interests. They may always be used as instruments to oppress industry.

A commodity set apart as money has a great advantage over all other sorts of wealth. It is wealth generalized, capable of being readily converted into anything produced by the industries of the nations using that money. It is therefore sought by every one more than any special sort of wealth.

But even if this were not true, bimetallism would prove a will-o'-the-wisp to the financial reformer. Past experience seems to indicate the impracticability of keeping in circulation two distinct money metals, each based on its own intrinsic value; and making one the standard and allowing the other to be redeemed in it is childishly foolish. A paper dollar redeemed in gold is more serviceable than a silver dollar redeemed in gold, for the latter as well as the former becomes a mere circulating obligation, and the silver in it is of no more value than a grain of sand. Of course when it ceases to be a dollar and becomes grains of silver the value returns.

But to return to bimetallism proper or the use of both gold and silver on an exactly equal footing. The same influences do not always control the prices of gold and silver, and there is no assurance that fixed weights of each will long remain of the same relative value. As a matter of fact, they are both constantly fluctuating in price. The principle, that of two metals circulating together as money, that one whose intrinsic value is less as compared with its face value, will drive out of circulation that whose intrinsic value is greater as compared with its face value, is pretty well established under the name of Gresham's law. The expense of coinage has a slight tendency to modify the law, as have also other considerations, but still its manifestations are very marked. Silver was twice almost entirely driven from circulation because of its relatively high intrinsic value, and if silver was coined to-morrow and made legal-tender for any amount at a ratio of sixteen to one, there would not be a gold piece in circulation in six months. Under Gresham's law, the dual gold and silver standard would produce fluctuation of standard and value as surely as would a gold or silver standard. But one metal would circulate at a time. When the demand for that metal for coinage, or in the arts, or both, or when other causes advanced the price of the metal, relative to its money value, beyond the price of the other money metal, the valuable metal would be melted down and sold as bullion and the cheaper metal would come into circulation. By the time the greater portion of the cheaper metal had gotten into the circulation, the greater portion of the dearer metal would have been driven out. The circulation would still be contracted to the volume of practically but one of the money metals. This would entail a heavy loss in coining and recoining and in no way relieve contraction of the currency.

Of course where one metal is subsidiary to the other, the law has no effect.

It might be possible to establish bimetallism with a composite standard, and secure a volume of metallic currency somewhat adequate to the needs of the nation, but that would come so near a sane solution of a bimetallic system that it would be scouted by every "statesman" and "practical financier" in the world. And it would still be a metallic currency incapable of responding to the laws of supply and demand, or of maintaining a fixed standard. A scientific money cannot be based on one or two metals or other commodities.

Fluctuation of the money volume or standard has many attendant evils.

There is little need to argue that contraction and expansion of the currency, with a consequent lowering or raising of the standard, is undesirable, but I will point out a few of the attendant evils. Contraction means high-priced money, a gain to the creditor at the expense of the debtor, an increase of interest and all unearned incomes at the expense of the producing masses, an easily monopolized medium of exchange. It means a fall in prices with a consequent diminution of the returns to active business men, a curtailing of business enterprises and an overstocking of the labor market. An overstocked labor market means lower wages. Give a greater return to the money-lender, increase unearned incomes, and the producer must get less. A relative rising of the money standard produces a like result. A dollar which remains stationary in value while the price of each commodity which the dollar stands for falls twenty per cent, gives to the holder of that dollar, or to him who collects a debt measured in that dollar, the same advantage as though the dollar had appreciated a like amount. The reason is, that dollars measure relative, not absolute values, and all debts are ultimately paid in goods. The application of improved methods and machinery to production brings down the price of commodities, and unless we would give all the benefit of this improvement to the money-lender, we must have a responsive money standard. But the tendency of all money standards so far devised, has been toward positive appreciation. It is necessarily so with any single commodity standard. And the rise in wages consequent to such appreciation is more than counterbalanced by the contraction of business and the consequent overstocking of the "labor market." Periods of currency contraction have always been times of distress among laborers.

I am aware that Economist Walker has given statements to prove that a panic has always occurred at times of currency inflation. But if his statements are examined it will be found that the panic really occurred during a time of contraction from a larger volume of currency to a smaller. The panic culminated in severe contraction.

Inflation of the currency of a country or a lowering of the standard of value below the average price of commodities, gives an advantage to the debtor at the expense of the creditor. It artificially enhances prices, gives an unearned profit to the holders of goods, and hence stimulates speculation. Stocks of metallic money or money metals share this appreciation, to the benefit of the moneyed class. Inflation lowers the real income of persons of fixed salaries, and has a detrimental influence on business morals and methods. On the other hand, inflation increases the volume of business, increases the demand for labor, and produces a temporary prosperity among the laboring masses. At least, such, according to Ramsey, is the lesson taught by the inflation of the Revolutionary period, and we yet hear laborers speak of the good times following the inflation of the civil war. But I am inclined to think that the cause was deeper, and inflation beyond business need is a dangerous experiment.

CHAPTER XXVI.

To disarm is better than to stab.

As we have seen, our currency is unstable as to volume and standard, easily manipulated and controlled,
and if placed on a metallic basis will be contracted to
a volume entirely inadequate to do the business of the
country. The bank is a ready instrument for controlling the currency. It dictates our financial policies, and
levies tribute on our commerce. Every law connected
with currency and banking has been dictated by those
whose business it is to exact interest for themselves
and others. The mechanism of all others which affects
the whole people, from the lad who collects and sells a
dozen of eggs to the manufacturer who disposes of a
steamship, is more closely monopolized than the tele-
phone or the electric light. Although the idea of money
is as old as history, the patent on the machine has
never expired. It still pays a royalty to private individuals, and the power for harm wielded by the manipulators of the financial machine is eloquently testified
by smokeless chimneys, silent wheels and starving
workmen.

The railway and the telegraph are powerful monop-
olies, often used by those who control them for oppression and extortion, but the banking monopoly is a
hundred times more dangerous. It controls all monop-

168

olies, presides at the exchange of all commodities, and fixes the measure by which all wealth shall be meted out. You may as well farm out the police power as the banking privilege. Indeed, in the present state of the world, cunning and trickery acting through powerful organizations are infinitely more oppressive than mere brute force. There is little doubt but that mere popular sentiment would be ample protection against forcible wrong, while the people are entirely at the mercy of the cunning and unscrupulous. Popular sentiment is too obtuse to see the wrong in its true enormity. It is the business of economists and sociologists to point it out, and of the people through their government to abate it. "He who controls the dollar rules the earth," is literally true in the nineteenth century, and the people must gain control over the institutions which control the dollar, or lose their hold on government. It might be well said, "Let me make your currency laws and let him who will command your armies."

It will not be denied that as an instrument for transferring and canceling obligations, the banking system has great efficiency. If it were open to the whole people its efficiency would be twice as great. It would be capable of being made well nigh perfect. But is it now a safe institution to control the finances of the people? We have seen that the currency system has a very questionable and crude foundation. The banks necessarily share its weakness. Indeed they are part of the currency system.

But the banks have weaknesses of their own. The banking capital of the United States was, in a recent report of the Comptroller of the Currency, estimated at $1,091,793,959, the total resources of the banks at $7,088,571,817, the cash assets at $515,987,740 and the deposits, individual and saving, at $4,535,908,584. Thus while the ultimate resources of the banks are apparently sufficient to meet all demands, the cash assets are at any time sufficient to meet but a very small percentage of the demand which may be made by depositors alone. The banks are literally kept up by the assumption that most depositors will not claim their own.

They are absolutely dependent on confidence and credit. But in time of panic, the consciousness of this very weakness, if not the greed for gain, would oblige them to hoard all of the ready cash which they can get their hands on, and keep it from the channels of business until the storm blows over. Depositors withdraw from the banks and the banks in turn withdraw from business the much needed money until nothing is left to carry on exchanges. To take money from business in time of panic is like drawing blood from a consumptive when he needs every spark of energy to keep up his vitality. And the banks are peculiarly well situated to make the draft. That they do so in times of distress goes without saying. During the last panic, it is true, the bulk of the banks saved themselves, but at what a terrible expense to all other lines of business! As is always the case, when bank favors were most wanted they were not to be had. If depositors had become as much frightened as the banks themselves, there would not have been an open bank in the country at the end of 1893. "In its incipiency it was a bankers' panic," says a prominent railway president.

Yet ours is such an excellent "sound" money system! A banking system so stable as to arouse the ecstatic admiration of statesmen and financiers! The stability is demonstrated when the "system" succeeds in bankrupting the industries of the country instead of bankrupting itself.

It is quite true that from the standpoint of the financier, which to-day is but another and milder term for Shylock, the present banking system is perfection itself. It may issue as much or as little money as it may choose, limited only by the value of the United States bonds it can control. It can collect interest on the *basis* of this money, and also the money itself. It can create a gold panic at will. It can make money stringency by the turn of its hand. If laws do not suit it, it can thrust its long tentacles into the purse of the nation and threaten to sap it dry unless laws are made to suit. In the inelegant but expressive slang of the times, the banking and financial combination has the country by the leg and can make its tackling effective whenever it chooses. Is that a safe state of affairs for a great nation?

There is no solution of the interest or currency question while the currency system remains in private hands, manipulated for private gain. The banks are a most important part of the currency system. They are now the great instruments of the usurer. To make the usurer harmless these instruments must be taken from him. Thé only way to make the currency system serve the whole people is through the people's government. We must have a national banking system in fact as well as in name. Constitutional powers are ample for the accomplishment of this end.

The power to issue money has always been considered the peculiar prerogative of the people through the state, and there is no generic difference in coining or issuing money and establishing a banking system. A bank is an instrument of exchange as truly as a dollar is. Any power which has the control of one must have the control of the other.

CHAPTER XXVII.

The ground is cleared; what shall the structure be?

A REAL national banking system must be under the direct control of the government. All past experience has demonstrated that where individuals were given irresponsible power, they used it to advance their own ends, regardless of others. However crude and imperfect the control which the people can exact over an institution, it makes that institution more responsive to the people's interests than would be any institution wholly unaccountable to the people. These principles are founded necessarily on the selfish motives which dominate the business world. Give men a privilege and they soon claim it as a right. The elected chief becomes in the next generation the hereditary dictator, and in the third, the king by the grace of God, the absolute dispenser of the happiness of his "subjects." Just so with the financial ruler. He secures the franchise by any means, then he owns it as a vested right, then he alone has a right to say how it is administered.

The problem is to institute a banking system controlled through the government of the people. It is very simple. No bonds need be issued, nor debts contracted beyond the rental of offices and the salaries of clerks, and it is to be hoped that the former item would not long remain formidable. It is not necessary to do a banking business for the people, but the banking business of the people. The gross product of wealth for each year, while in the process of exchange, will serve as ample capital.*

* It must be borne in mind that all goods offered for sale must be weighed, measured and stored at present, and I propose no additional trouble.

172

1. Let banking houses be established by the government in every city, village and hamlet where there is a market for produce sufficient to warrant such an institution.

2. When any commodity for which there is a market, is ready to sell, let the owner, if he so desires, place such commodity in a warehouse, private, corporate or public, bonded and registered for the purpose, and receive therefor warehouse receipts.

3. The receipts should be issued by weighers, gaugers or inspectors, elected by the electors of the county, state or municipality in which they serve, but directly responsible to the head of the United States banks. The warehouse-owner or his agent, where the warehouse is private or quasi-public, may countersign the receipts. If the warehouse be public this should be done by the officer in charge.

4. On the presentation at the government bank of these receipts, and the giving of additional security to the extent of the market value of the goods therein represented, let the bank take up the receipts and give the presenter credit on the books of the bank for the market price of the goods stored, or issue him full legal-tender paper money for that amount.

5. In the bond of security required of the person thus opening a bank account, let it be required that he sell the goods or redeem the warehouse receipts for cash within eighteen months from the original storing of the goods, under penalty of forfeiting his goods and bond or sufficient thereof to indemnify the bank for the credit advanced. The bank should be required to turn over the warehouse receipts to the owner or his assignee on tender of the amount of legal tender, or cancellation of the credit issued thereon.

6. Let the warehouses be open to all classes of goods as at present, with the requirement that they be for sale in good faith, and let the banks issue credit or paper on goods for sale alone.

7. Let bonds, stocks and evidences of credit be taken as collateral on the security bond, but let no money be issued on anything except real tangible wealth at its market price at the place where the bank is located.

8. Let arrangements be made by which goods in transit may become the basis of bank credit or currency.

9. Let all branches of legitimate banking be conducted by the government at cost, and the revenue be raised by charging a small commission on transactions.

10. Let actual currency deposits be also made the basis of bank credit.

11. Let the banking department have full charge of the issue of all money. Let the coinage of all metallic money cease and let gold and silver cease to be legal-tender for debts.

12. Let the precious metals, so-called, be placed on exactly the same footing as other goods and allowed to serve as a basis for the issue of legal-tender paper.

13. Repeal all of the present laws relating to banking and currency, except such as may be applied to the new system, prohibit private individuals from entering the banking business, and establish the proposed system by appropriate legislation as to details.

This is a plan for a scientific currency and banking system. I do not insist on details, but I do insist on the underlying principles: that all money shall be issued on wealth in the process of exchange, that the actual circulating medium or money taken shall have no intrinsic value, or as small an intrinsic value as may be, and that the entire currency and banking system shall be under the direct control of the government. The end and aim of currency is to facilitate exchanges.

Such a currency and banking system would strike at the very foundation of usury, and would cut borrowing down to a very small volume indeed.

This would entirely relieve the necessity of borrowing in commercial transactions. The discounts on bills of credit would be immaterial. There could be no cornering of the currency and no paucity of medium of exchange.

But if money could be hoarded there would still be a temptation to hoard it, and to make the remedy against usury perfect, the currency must be made of an especial sort.

1. Present interest laws should be repealed and interest-taking made punishable as a crime. The punishment should not be made unnecessarily severe, nothing more than the forfeiture of the entire debt, one-half to go to the informer.

2. Loans between private parties should be made absolutely uncollectible except where made through government banks, and it should be made the duty of the bank officer to see that the whole amount of money for which security was given should be paid over to the borrower. Soliciting back any portion should be considered and punished as fraud or getting money under false pretenses.

3. Issue money stamped conspicuously quite across the face with the date of expiration, year and month, issue twice a year and date on any two months of the year, six months apart, and make such money current but eighteen months from the date of issue.

4. After an issue had become non-current, allow it to be redeemed at par at the bank for a few days, with the requirement that all money presented for redemption should be left in the bank until the entire period for the redemption of the issue had expired. The redemption would be made in current notes running eighteen months longer.

5. Where current notes of the face value of a certain sum, say two hundred dollars, were presented for redemption by any one person, let him be charged a percentage for the redemption on the amount of the notes presented in excess of his average deposit or credit at the bank during the time for which the notes were current. Charge all depositors a percentage on their cash deposits in excess of the credit issued to them on goods.

6. After the few days were past during which the currency should be redeemed at the bank at par, let a uniform percentage equal to the average deterioration of wealth be charged for redemption, until the non-current notes should become worthless.*

*While I recommend most emphatically that the banking system be owned by the government, the proposed plan may readily be applied to the present banking system.

This would make it no more desirable to hoard money than to hoard goods, and would force those who received surplus incomes to invest them in real wealth or to loan them without interest, for use in production.

CHAPTER XXVIII.

In the light of what is, we may discover what should be.

THE practical working of the system would be very simple. We will suppose, for illustration, that a bank is established in a small town surrounded by an agricultural population and carrying on within its borders some manufacturing business. The town, of course, is supplied with stores, so that nearly all functions of the commercial bank will be brought into requisition

To begin with the simplest, we will suppose that the farmer wants to market his crop. He may go into the city and sell it as he does now to the local dealer, and then it would be the local dealer who would do business with the bank. He may want to hold a portion of the crop for a few days and borrow on it. He may want to take his time at selling it. In that case he will take a load or several loads into town and, as farmers do at present, store it in one of the grain warehouses, taking warehouse receipts therefor. He will then file a bond with the bank and present these warehouse receipts or a portion of them and receive bank credit or money therefor. With this money he will go to the merchant and purchase a portion of his year's supplies. The merchant will take the money to the miller and lay in a stock of flour for his town customers, the miller will

177

go to the farmer and purchase his wheat to keep up his stock of flour and the farmer will use the money to redeem the warehouse receipts and release his wheat. The money which the farmer has issued on his wheat might serve as the medium of a dozen transactions. The farmer might use it to pay the blacksmith, who might use it to pay his helper, who might pay it to the doctor, who might pay it to a music teacher or artist, who might pay it to a merchant, who might pay it to a manufacturer of shoes, who might pay it to a lumber man, etc., until that or other money would be paid again to the bank to redeem the warehouse receipts, in pledge and release the wheat from the warehouse.

Just in the same way the manufacturer of furniture, let us say, might store his goods, obtain receipts, and presenting them with his bond at the bank, obtain money with which he might buy raw material, pay wages to workmen, etc., etc., until such time as he might sell his goods, when he would use the money realized from the sale, or a portion of it, to redeem the warehouse receipts and turn over the furniture to the purchaser.

The miller might store his flour and get money on it to buy wheat, pay for his barrels and his hands and keep his mill going. The necessity for redeeming his pledges would, of course, make it incumbent on him, as well as all the rest, to sell his goods as rapidly and as advantageously as possible, but while he had an equivalent in value he would have no trouble in purchasing what he wanted or of using it as live capital to carry on his business.

The merchant might store his surplus stock, draw it from the warehouse in consignments, and use the money issued upon it to meet current expenses and keep up stock. The miner might store his product and have money issued upon it to keep up his business while he negotiated its sale. In any branch of business there would be no such thing as a scarcity of money. As soon as goods were produced and placed in the common stock their representative currency would be placed in circulation. Each dollar's worth added to supply would be a dollar added to effective demand for goods of some sort. Commerce could never become stagnant for want of a medium of exchange. This town would be a typical town.

There could never be too much money, for every dollar added to the currency would presuppose a dollar's worth of wealth placed publicly on sale. According to the unswerving law of economics there is a supply for every demand, and the dollar founded on demand would find its supply.

In the natural course of business the money would be first issued to the original producer. He would pay it to those who would place it in the hands of the retailer. The retailer must turn it over to the wholesaler and the wholesaler pay it back to the original producer to go back to the bank of issue.

There never yet was a proposition put forward which did not meet with objections, and this is no exception. Let us look over a few of the more prominent.

There is not an intelligent financier living who will not admit that a currency founded on value to the full extent of the issue, and perfectly and quickly convertible, is an ideal currency. The proposed currency has both of these qualities. It may be said that the proposed currency is not convertible into gold. I answer that the whole stock of gold in the country is at your disposal if you have the wherewithal to purchase it. Is more to be had at present? And upon what is based the superstition that gold is the only sort of wealth which will cancel a debt? Simply on the traditional worship of the golden calf. Gold will cease to be legal tender as soon as it loses that quality by law. As soon as people become enlightened enough to realize that a dollar's worth of iron is really as valuable as a dollar's worth of gold, they will see without difficulty that a note redeemed in any sort of wealth which the owner may choose, including gold, is as truly a redeemable note as a note redeemable in gold alone.

If under the proposed currency you want to ship gold to other lands in exchange for goods, it will be much easier than now to obtain it. It will be more plentiful and cheap, for the great demand for it as a money metal will have passed away. As was said above, there are all other sorts of wealth into which one may convert his currency, and currency converted into gold or any other form of wealth you may choose, is certainly more truly convertible than currency convertible into gold alone. If quick and certain convertibility is a quality of good currency, then the proposed currency is superior to any currency yet known. The fixed security, as well as the deposit of wealth on which the money is founded, makes the convertibility absolutely certain. A dollar is redeemed every time it is accepted for goods.

Paper money absolutely secured is an ideal currency. Thus economists agree that certificates issued on coin are superior to any other form of currency, for they are quickly redeemable, perfectly secure, more easily handled and in every way better adapted to the processes of exchange, than is the coin on which they are founded. Wealth certificates are necessarily superior to coin certificates. They have all the convenience and double the security of the former and are issued on such a principle as to make them available for any one who wishes to exchange wealth. In fact they have greater convenience than have any coin certificates, for a coin certificate without the legal-tender quality would be good for the coin only, while the proposed wealth certificate would be good for any sort of wealth in the market. If the coin certificates are made legal-tender, then they are simply limited wealth certificates used as money, and the coin becomes simply stored wealth, differing in no way from the wealth proposed as a basis of currency.

Up to a volume equal to the amount of wealth at any time offered for sale, the proposed currency would be perfectly stable in unit of value. It is commodity money, or money founded on commodities small in value as compared with the wealth represented, whose unit value is affected by change of volume. Above the volume of wealth offered for sale, this currency volume could not go.

Economists have long given unsatisfactory explanations of the considerations on which depend the value of the money unit, and a few recent writers in trying to make it clear, seem to have involved the matter more than ever.

The considerations governing the value of the money unit depend, to some extent, on the sort of money which is under consideration. If the money be credit money, the value of the unit depends on the volume issued, above or below the normal amount. Thus, if we conceive the stock of wealth to be twenty sheep and conceive these sheep to be represented by one hundred dollars, each dollar will represent one-fifth of a sheep. Now let the currency be swelled in volume so that the sheep are represented by two hundred dollars and each dollar will represent but one-tenth of the value of a sheep. Let the currency be contracted to fifty dollars and each dollar will represent the value of two-fifths of a sheep. Here we presuppose that the credit back of the currency is good and that the money token has no intrinsic value. In a state like that the salable wealth in the country would be represented by the dollars in the currency of the country and the value of the dollar would be inversely proportional to the volume in circulation. Up to the limit of the credit the currency unit could in no way be affected by the relation of the credit to the volume. Every dollar of the currency would be a demand for the part of the salable wealth represented by a fraction with one as a numerator and the number representing the volume of the currency as the denominator. It will be readily seen that any increase in the denominator will decrease the value of the fraction or the money unit, and any decrease in the denominator or contraction of volume of currency must increase the value of the fraction or the value of the money unit.* What is the actual volume of money, is not of such importance, provided the money be allowed to circulate freely and the volume remains

*It is true that with a mixed or heterogeneous currency like that in use at present, referred to a single commodity as a measure of value, the value of that commodity, other things being equal, determines the prices of other commodities. But the value of the money commodity is determined largely by supply and demand, and anything which tends to diminish the demand for the money commodity tends to lower its (exchange) value. It is obvious that in the proportion which other sorts of money are used the demand for the standard commodity as a money commodity is lessened, and and its value decreased. Then, here too, the law of inverse proportion between volume and unit value tends to hold good.

constant. If we did a strictly cash business it would
not make much difference if horses sold for ten dollars
each and wheat for twenty cents per bushel, provided
everything else sold accordingly cheap and these prices
did not fluctuate.

With a commodity money, or a money token of
intrinsic, equal to its face value, the volume of the currency has but a secondary consideration in determining the value of the money unit. The value of the
money unit depends on the price of the commodity of
which the money is made, provided always that there
is free coinage or manufacture into money in unlimited
quantities of the commodity used as money. Thus, if
the gold dollar becomes more valuable than the gold
in the dollar, the holders of gold bullion will have it
converted into dollars until the increased volume of
the dollars will bring their face value down to their intrinsic value. Conversely, if the weight of gold in the
dollar would sell for more as bullion than it would buy
as a dollar, the gold coins would be melted down or the
gold bullion held from coining until the scarcity of
coin or metal to coin would bring the face value up to
the intrinsic value of the dollar. If this were not so,
where gold alone is money we should find a discrepancy
between the face and the intrinsic value of the coin,
which is contrary to experience.

The value, then, of the gold money unit, or any other
commodity money unit, must depend on the price of
the metal or the commodity, independent of its money
value. This price depends almost entirely on the supply and the demand of the metal. Not the demand for
coining alone, but the demand for coining and all
other purposes.

Volume of circulation plays a secondary part. Where
there is an increase in the volume of circulation, the
supply of gold remaining constant, the demand for
gold to coin must increase the price of the metal and
therefore appreciate the money unit. A demand for
gold in the arts would have the same effect. Now if
gold were the only money in circulation and the wealth
of the country remained constant, the increase in the

volume of the currency, which would increase the demand for and therefore the price of gold, would in a measure have a compensating effect by decreasing the purchasing power of the dollar. But this effect must be rather theoretical than practical, for as a matter of fact, no gold could actually be drawn from the arts while it was more valuable for use there than as a money metal, and the currency volume could not be increased with a falling money unit, except where the supply of gold for all purposes outran the demand so as to make its price fall everywhere.

Then again, the wealth of the country is not constant, but increases so rapidly, as we have seen, that the amount of gold which may be used for coinage cannot possibly keep the volume of money proportionally as large as at present. On a gold basis the relative volume must ever decrease and the unit appreciate.

Then our mixed currency, some founded on one principle and some on another, prevents any uniform manifestations of monetary economic law.

We see, then, that the value of the money unit depends, in a commodity or metallic currency, on the market value of the metal of which the money is composed; that the laws of supply and demand keep the money value and the market value of the metal equal, or nearly so, and that the volume of money has but a secondary effect on the value of the unit.*

The larger the volume of other currency founded on the standard money metal and redeemable therein, the more fluctuating the demand for the standard money metal and the more variable the money unit. Auxiliary currency is really but a tool for manipulating the standard currency, and if the shock of contraction could be

---

*Strictly speaking, no money while it is being used as money, has intrinsic value. When being used in exchange it is a representative of wealth to its face value. A gold dollar cannot in the same transaction have a representative value of a dollar and also an intrinsic value of a dollar, for if it had we could get two dollars in wealth for every dollar in gold But that is contrary to experience. We can trade the metal in the dollar for a dollar's worth of goods (that is barter), or the dollar itself for a dollar's worth of goods, but we cannot buy wealth for both the dollar and the metal in the dollar. When used as a dollar the gold in the gold dollar is not worth a rap. Intrinsic value is entirely sequestered and becomes available only when the dollar becomes a commodity. That is ones reason why a metallic currency is so extravagant. The metal in the money lose all intrinsic value while it is used as money, and that amount of wealth is practically subtracted from the nation's assets.

once gotten over, the country would be better off in all respects with but the standard currency, without the mongrel auxiliaries, useful only in manipulating the currency.

It may be said that the silver coins in present use do not accord with the laws enunciated above. Their money value is greater than their intrinsic value; their money value and the price of the metal have nothing to do with each other. It must be borne in mind that silver is not, and has not for the last twenty years, been a money metal in the strict sense of the word. The coinage of silver was not free but arbitrary, and therefore the face value of silver coins was allowed to appreciate above the bullion value without its being possible to offer enough for coinage to raise the price of the white metal to a like standard. As a matter of fact, silver money as at present used is mere credit money, differing in no essential manner from paper money. Its standard is gold, and the fixing of the price of gold fixes the value of the silver as well as the paper dollar.

Whether a money must have intrinsic value, depends on one's definition of money. A recent writer labors to show that all money must have intrinsic value and goes on until he finally makes money synonymous with wealth. Carry his system to its logical conclusion, and all wealth would become money. This is absurd and leads only to a confusion of terms. Scientific money is a mere medium of exchange, a mere token of title, and cannot be identical with the wealth exchanged. It is in this sense, at least, which I wish to use the term in the following pages. I wish to make as sharp a distinction between the currency and the wealth which the currency is founded upon as I would between a title deed and the land, the ownership of which was indicated by the title deed. The one is the instrument of transfer and the other the thing transferred. There is no advantage in jumbling them together and calling them both by the same name. It is positively erroneous. Money is at least but a generalized claim for wealth to the amount of the current value represented by the

A BREED OF BARREN METAL

former. You cannot generalize any particular sort of wealth as such. By money I mean the money token, nothing more. Currency I use in a broader sense as any contrivance used in exchange, including banking systems. The token need not have value, but must represent value.

It may be said that the proposed money is paper money and therefore not desirable. If you mean paper credit money or fiat money, in the sense in which it has been heretofore known, then the statement is incorrect, but we have had no experience which would warrant us in condemning even this. There has never been a pure paper or credit currency in any nation. All systems of paper currency so far established have been exploited side by side with gold and silver. It has always been the established policy of the most influential financiers to do all in their power to discredit and depreciate the paper currency and finally turn to specie. The paper money in our currency system has been little else, in times of peace, than a tool with which to manipulate the standard metallic currency.

Adam Smith, who is considered an authority on such matters, did not distrust paper money because it was paper. He writes:

"A paper money consisting in bank-notes, issued by people of undoubted credit, payable upon demand without any condition, and, in fact, always readily paid as soon as presented, is, in every respect, equal in value to gold and silver money; since the gold and silver money can at any time be had for it. Whatever is either bought or sold for such paper, must necessarily be bought or sold as cheap as it could have been for gold and silver."

While the gold and silver money could not be had for the proposed currency, gold and silver or any other sort of wealth may be had for it, and it would therefore be better than the best of bank notes.

Let us briefly review the historical instances in which paper money is said to have proved a failure and see whether this paper money had much in common with the proposed currency. Continental money is one of the first object lessons which the advocate of metallic money will point out for the benefit of his paper currency brethren. The Continental currency was issued by a nation without autonomy, without credit, without taxing power; a nation whose very existence depended on the desperate chance of a few scattered colonists defeating in war the greatest power of Europe. It was issued as a credit money; no means were provided for its redemption; yet it floated at par to a considerable volume. While the fortunes of the war and the prospects of a nation were even fair the money circulated, and was finally made worthless only by over-issue, counterfeiting and threatened repudiation. We can then gather from this instance nothing as to the stability of a money founded on good credit, nor a paper currency founded on wealth to double its face value. The issue of paper money at the time of the Revolution was simply a means of obtaining a loan or rather a contribution, and can in no way be compared with the issue of money for the ordinary needs of business.

The issue of greenbacks at the time of the civil war, as well as the issue of other circulating paper was made simply on the general credit of the nation, at that time menaced by the most formidable rebellion of history. That it depreciated is not surprising. That it circulated at all is far more so. Then in the issue of paper money in war time there was no criterion as to what amount was necessary to carry on the business of the nation. The channels of trade were relatively overloaded. The paper money itself was discriminated against by law, in not being made receivable for all debts, and it was exploited by the side of metallic money which was cornered by financiers and used as an instrument to depreciate the paper currency for the purpose of speculation. From the facts surrounding the civil-war issue of paper money we are warranted in concluding that the money was as good as the credit of the issuing nation. We have good reason to believe that had the credit been perfect, the money would not have depreciated until over-issued, especially if it had full legal-tender qualities and was not subject to the manipulations of speculating financiers.

Confederate money was founded on poor credit, degenerating into none at all, was over-issued and counterfeited, and hence is no criterion of the feasibility of a credit currency founded on good credit and issued up to the limit required for business.

The French assignats come the nearest to being an example of pure credit money, but the inadequate measures taken for their redemption, the want of a guide as to the bulk which was needed in the country, and the very precarious state of the issuing nation all conspired to place the system in bad repute. In this instance, as in all others, conspiracy of the financiers of the world against any form of currency which threatened to dethrone the dominating influence of the precious metals, had much to do with the overthrow of the French currency system Counterfeiting also played a very important part. The Continental currency of America, as well as the French assignats, was counterfeited to an enormous extent, and this had much to do with the final worthlessness of both. Yet they both served their purpose. Without the Continental money the American Revolution could never have been brought to a successful issue. Without the forced loans from the nobles which were obtained largely through the assignats, it is doubtful if revolutionary France could have kept feudal and monarchical Europe at bay. The effect of decentralization produced by the assignats was a boon to France. The result of the War of the Rebellion was largely affected by the issue of paper currency. But if all of these were absolute failures that fact would not prove the impracticability of the proposed currency.

This all goes to show that the fact of a currency being paper is nothing against it, as the weakness of all former issues of paper currency is readily explainable on the ground of the weakness of the principle on which the issue rested. A paper currency resting on a sound basis may be the best currency in the world.

If paper currency is so useless, why is it that at times of greatest national hazard it is always resorted to? If metallic money is so much more stable, why did we not stick to it in our struggle for national life and later for national existence? The fact is, metallic currency proved itself inadequate before either struggle had fairly begun, and the much abused paper was relied upon to tide over a crisis where commodity currency failed.

Other countries have had the same experience. England had to abandon specie in her Napoleonic wars and paper currency was her only refuge. It was not the first time that she had to resort to like measures. The French Republic had often the same experience. In time of desperate war Holland adopted much the same plan as I propose. If the history of finance has taught one lesson more plainly than all others, it is that a metallic currency is utterly unreliable in times of national calamity. The phenomenon is easily explained. Metallic money is the safest and easiest sort of property to hoard. All who can, will in times of national or financial disaster turn as much as possible of their surplus property into gold and silver. There is always sufficient surplus wealth to appropriate all of the money metal in circulation, and in times of pressure a nation with a metallic currency is left practically without a circulating medium. That which escapes the miser gets into the hands of the speculator, who uses it for extortion and places himself in such a position that when peace comes he may reap a harvest.

But we do not have to wait for times of war to show the inconsistency of metallic money advocates. The apostles of the gold mania, the very bankers who a year or more ago clamored so loudly for gold, have lately put forward a plan to supply the nation with a paper currency. Their idea seems to be that paper is very good if the issue is under their control and made in such a manner as to allow them to reap a substantial benefit. Let metallic money advocates point out a country where metallic currency has served all the needs of trade in times of peace, or met the exigencies of war, and we will then be ready to listen to their tirades against paper money because it is paper money. But there is paper money and paper money, and because wild schemes of unlimited or arbitrary issue of paper money are subject to criticism, is no reason why we should doubt paper money fully secured and issued only as demanded by the exigencies of trade.

The money proposed by the bankers is quite as unreliable as any fiat issue yet concocted. What has the amount of banking capital to do with the volume of exchanges carried on within a year? One is no criterion of the other.

The supply of the proposed currency must always equal the demand. Supply and demand in general are always equal. Any dollar's worth of wealth may have issued thereon a dollar for effective demand, and, therefore, one cannot imagine a case where a demand for money will go unsupplied. The currency is issued on wealth intended to supply a demand only, and therefore there can be no redundancy of currency. The criterion is whether the wealth is for sale. After money has performed its function of exchanging wealth it is withdrawn from the circulation simultaneously with the wealth which it exchanges, making over-issue impossible. No commodity money can be regulated in that way, for the value of one or two commodities must be less than the value of all other commodities, and as all other commodities must have that one commodity for a representative, the demand must necessarily exceed the supply. This is still true even when we consider that demand for currency is made only by commodities for sale. It does not help matters to say that the whole hoard of gold or silver is set over against the commodities being exchanged at any one time If a whole year's product is not exchanged at one time, neither is the whole volume of money mediating, or capable of mediating exchanges at any one time   The bulk of the currency is usually out of active use, just as the bulk of goods is out of active trade. A single commodity money must necessarily be subject to all the fluctuations of the commodity which composes it.

All commodities taken together can never fluctuate in exchange value. They can change in respect to one thing only: the labor by which they are produced. However much more effective that labor may become, the same relative productive effort will always command from nature the same relative remuneration. It is only where the idler steps in and takes a portion of the production, that this law is modified. A money issue founded on all commodities can, then, never fluctuate. There could no more be too great a volume of such a currency than there can be too great a volume of commodities in general. Every dollar would represent a dollar's worth of wealth capable of being purchased by that dollar.

CHAPTER XXIX.

The scales of justice are balanced alike for all.

THERE is no difficulty about establishing the proposed money standard. Gold is not properly a standard a t all, for it is subject to fluctuation. The very fact that it is used as money makes its price more unstable than it otherwise would be. The supply is limited. The volume of the metal produced in any one year is insignificant as compared with the volume in actual existence, so that the price is regulated scarcely at all by the cost of production and depends almost totally on supply and demand. This is very precarious. The supply of gold is ridiculously insufficient, as we have seen, to carry on exchanges, and hence the unsatisfied demand must make it tend rapidly upward until there is some new deposit discovered, when the fall is sharp. In times of pressure this fluctuation is accentuated by speculative manipulation, when the price of the metal often changes several hundredths in a few days. Thus debts are paid by a different standard from that by which they are contracted. It will take more iron or coal or wool to pay a debt, than could be purchased for the money borrowed at the time at which it was borrowed The creditor is made a present of the amount of the gold appreciation, a present which for the past twenty years amounted, on an average, to more than two per cent per annum.

While there may be a general fluctuation in money prices, there can be no general fluctuation in value.

191

Where the money prices of all staple articles fall, the
fact is that values remain the same and the money ap-
preciates. Values being relative, a general rise or fall
involves a contradiction. There can be no general rise
or fall in values as compared with any except the labor
standard. If by the use of machinery all articles can
be produced with less labor, that does not change their
relation to each other and hence their value. For this
reason, the gold advocates who claim that gold remains
stationary in value while all other articles depreciate,
make a statement which is misleading if not absolutely
false. Probably it takes no more labor to produce an
ounce of gold now than it did twenty years ago, but
the labor now required to produce an ounce of gold is
certainly greater in comparison with the labor required
to produce a bushel of wheat, or a pair of shoes, or a
coat, or a chair, or a carriage, than it was twenty years
ago. Hence gold is a false standard for gauging the obligations of debtor and creditor, for it is with wheat, or
shoes, or coats, or chairs, or carriages or some other article of production that debts are really paid.

It is asserted that contracts are made for payment
in gold and that hence the debtor but pays what he
really owes and has nothing to complain of. Such a
statement begs the question as to an injustice-working
currency. When one contracts a debt he intends to
contract to pay in principal just what he received. Our
currency systems are supposed to be especially designed
for the purpose of making contracts exact in this respect. If then by a stealthy appreciation of the currency, one is obliged to pay back more than he has
received, an injustice is done him. Even though what
he pays back costs him no greater labor effort than what
he received would have cost him at the time he received
it, he is done an injustice if what he is obliged to return is capable of supplying a greater number of wants,
or of purchasing a greater number of commodities than
what he received. Otherwise we must hold that the
creditor class is entitled to all return for increased
effectiveness of improved machinery. It is labor which
produces this effectiveness.

In fact gold is not a standard at all, but a sort of rough value denominator; and the exchanges made with gold on a gold basis are merely barter, a trading of commodities, an improvement on primitive barter only because of the fact that gold is more readily divisible and portable than the average commodity.

A fluctuating standard must be radically faulty. What would we say of a yardstick which would be thirty-six inches this year and thirty-nine the next; or a pound which would contain sixteen ounces this year and eighteen the next? Yet this is the exact position of the gold money standard. And the demand for gold has so outrun the supply that the gold dollar must ever remain the yardstick of increasing length, the pound of increasing weight. The scramble for the meager supply must forever push the price upward. Were gold not the tool of misers, usurers and monopolists, the whole product might be profitably used in the arts.

There is but one standard in the world which does not change and never can: a fixed quantity of the same quality of human effort. This standard has always determined the value of commodities in general and always will. Supply and demand may for a time affect the price of this or that commodity. (Where the demand is artificially regulated, as with gold, and the yearly product is insignificant as compared with that on hand, and the commodity itself is well nigh imperishable, demand may be the chief factor in determining the price.) But the average value and hence the price of any commodity is always dependent upon this principle: What relative amount of human effort of a certain quality did it take to produce that commodity as compared with that which you seek to exchange for it? Labor, the producer of all man-made wealth, is the only true measure of value. Human effort in general may be remunerated by greater results this year than last, but an hour's toil, whether of brain or hand, is an hour's toil this year, it was last and it will be as long as there is a toiling human. Man's judgment of the value of any article depends wholly on the effort required of him to secure that article. *Value to man is interpreted only in terms of toil.*

A writer says that labor is an act and not a quantity and can therefore never be used as a money standard. It would be interesting to know just what he means by the assertion. If it is that labor has not quantity, then the statement is manifestly false. Labor is nothing more nor less than productive effort. Productive effort has quantity as much as an electric current or the working force of a steam-engine. The effort of a day is greater in quantity than the effort of an hour, provided both efforts be of the same intensity. We measure the force of currents of electricity by results. We measure the working force of machines by results. We have a unit by which this is done, an actual tangible unit. Why can we not measure labor by results and in that way fix a unit which may serve as a measure of all results of labor? That would be a standard of value, a money standard. Whether in paying wages we pay for labor effort or for the goods, is entirely immaterial. Our judgment must in both cases be based on tangible results.

The same author gets the cart squarely before the horse when he says that it is the product which imparts value to labor. The fact is that labor imparts value to product. That which requires no labor, unless monopolized, never has any (exchangeable) value. That which requires labor has always more or less value and usually in direct proportion to the amount of labor expended. And as a matter of fact labor has value, for it is every day in practice exchanged for valuable substantial commodities. The hair-splitting is as unnecessary as erroneous. The author is simply misled by the fact that labor must necessarily be judged by results and rewarded accordingly. Land is not an example of value without labor, for it is the labor-produced civilization on earth which has given value (exchange) to land.

The ideal measure of value, as well as the ideal money standard, is a fixed quantity of human effort of a certain quality. This is really the yardstick by which the results of all human effort are measured. But we are too ignorant or unskillful to determine the quality of labor directly. We must therefore judge by results, and in this way we may closely approximate the ideal standard.

If there were no debtors and no creditors, there would be no necessity of fixing a tangible value for the labor-effort dollar. On a strictly cash system, the dollar may be made in terms the value of the result of an average hour's toil, and the market price of the commodity sought be left to determine what that value would be at any particular time. Currency loans would be paid in currency founded on wealth in the process of exchange. The actual money would be founded on the wealth which the debtor had added to the common stock, and it is with this that he would actually pay the creditor. The creditor would not receive that commodity, but the right to demand in the market any commodity which he chose, provided he paid the market price therefor. The creditor could demand neither wheat, nor corn, nor coal, nor iron, nor gold, directly from the debtor in any fixed quantities, but he could demand and secure that debtor's right to demand coal, or iron, or wheat, or gold, which at their market price would be represented by the number of dollars called for by the contract of indebtedness. If the same dollars paid for all commodities and all services at all times, it would be immaterial what the ultimate standard of the dollar was. Its value would be determined in every exchange.

But as there are debts, it would be necessary that the labor currency unit should correspond closely with the dollar now in use, or rather what the dollar should be. This, it was seen, would be the standard of 1873.

It takes a certain amount of labor to produce a dollar's worth of product in any and all industries. That neither debtor nor creditor should suffer by the proposed change of standard, we should determine what quantity of labor is on an average required to produce a dollar's worth of commodity, and we should make that quantity of labor the dollar.

There would be no difficulty in ascertaining what such quantity of labor is. The material is at hand. One can ascertain with accuracy the average amount of toil expended in producing a bushel of wheat, a case of shoes, a bolt of cloth, a ton of iron, a ton of coal, a plow, a locomotive, etc. Let the amount of toil required in each case be compared with the average price of the article in four or five markets, and from this it would be easy to determine the average amount of labor required to produce a dollar's worth of goods. This could be made the money standard and there would be no perceptible change. If it was thought best, the standard might be expressed in terms of product, but I deem labor terms the best for the expression of the standard. In practice, the standard would mean simply that the holder of a dollar is entitled to the product of a certain labor effort in the industry by which the commodity sought has been produced. What the amount of the product would be, would be determined by market price.

Buying and selling is merely a convenient mode of trading commodities for one another. The price of each fixes the price of the other, and the dollar is simply an expression of the relative amount of labor in each.

The standard once fixed, it need never change. No matter how much more productive an hour's toil may become, it becomes no more productive relatively than an other hour's toil of the same quality. All values are relative, all remuneration is relative. No matter how many more wants his toil will satisfy, any particular laborer would receive no more remuneration as compared with any like laborer than he does now. There would be so little long-time borrowing that the standard would give no material advantage to one over another.

Of course I recognize but two factors in production: the laborer and the spontaneous forces of nature. The latter claim no reward, the former should therefore have the whole product. Everybody capable of producing should labor, and the forces of nature should be so free from appropriation that they may distribute their favors alike to all. Distribution would soon become simple, and the relative remuneration of different laborers be the only question to be solved.

CHAPTER XXX.

We always find excuses for resisting disagreeable truths.

THE first objection which is naturally directed to the proposed currency, is that it is impracticable. Where are the specifications? What particular portion is impracticable? Is it the nationalization of the banks? The United States has found it practicable to establish postoffices in every village and hamlet in the land; to provide officers and employés therefor and to see that the business of carrying and distributing the mails is properly carried on. The establishment of a national bank system would be connected with little more expense; it would be little more complicated and it would be much more closely related to government functions than is the carrying of the mails. The United States now manufactures all of the currency, secures it and superintends its use. Under the proposed system, it would be asked to do no more in that respect. It has charge of all national bank accounts; it primarily issues all circulating notes. And for whose benefit? Largely for the benefit of private individuals. All that would be necessary in addition, is the local officers, with access to town, county and municipal records, so that the value of this or that security might easily be ascertained. The weighers and gaugers would be part of the local force.

Banks could not possibly be more numerous than

197

postoffices, and a government capable of directing the work of postmasters would be capable of directing the work of bankers.

The bank officers and post officers may be the same in small towns, and thus the expense of both services might be cut down. The same buildings, too, might be used for both concerns.

Nearly all commodities are now stored before selling, and the only additional requirement would be to have the warehouse bonded and registered.

All commodities must now be weighed, measured or estimated in some manner. The price of all commodities must now be ascertained before the commodities can be sold. The storing, weighing or gauging are not new requirements.

Deception in weighing, gauging or the issuing of receipts would be difficult, as the banker would be a check on the weigher or gauger and the weigher or gauger on the banker. The warehouse man would be a check on both and the owner of the produce on all. Collusion would not help, as the amount of the money issued must be returned before the produce could be taken from the warehouse or assigned, and there would therefore be no object in overestimating price or weight. The bond of security would make it utterly impossible that the public should be swindled or sustain any loss whatever. Officials without any pecuniary interest in falsifying would be less prone to dishonesty than the officials under the present system, who have a direct pecuniary interest in deceiving.

The currency would be issued on the price of the goods in the market at which the bank was located, in the quantities in which the goods were offered at the warehouse. Whether the price would be the wholesale or the retail price must depend entirely on the local market for the goods or their value at the banking town. Thus, there must in practice be a minimum amount of goods capable of being stored as a basis for a money issue. This must depend on the locality in which the bank operated. With agricultural customers, the minimum must be as low as twenty-five dollars or thereabout, while in a manufacturing community where establishments were large the limit might be much higher. The price would be reckoned on the lot of goods stored, whatever that may be. If wheat, why, wheat in the local market. If merchandise, stored by a local merchant, then the price of the goods would be the price of that bulk in the local market.

In practice, no less than the minimum amount of goods allowed to be stored as a basis of money issue by any individual, should be allowed released or assigned at any one time Thus the storing of goods would in no way interfere with their sale. A merchant might keep his surplus stock in the bonded warehouse and have money issued thereon and release it in installments as he wanted to sell it. So might a manufacturer. In that way nearly the whole volume of wealth offered for sale might serve as a basis for currency. Samples alone would be required in salerooms.

It is not necessary that every dollar's worth of goods *shall* actually be represented by a dollar in currency, but it is necessary that every dollar in goods *may* be represented by a dollar in currency if required for the purpose of exchange. Thus the volume would regulate itself, there would be no wealth to exchange which might not secure the dollar to exchange it, but if it could be exchanged without calling for the issue of currency, the exchange would be allowed.

The warehouse receipts would not be the sole security. The fixed bonds would be sufficient for that, so that the former could be regulated so as to be readily assignable and cause no friction in trade. They might be issued in duplicate, the copy being made assignable, in order to facilitate trade.

The security required could not be a deterrent to trade. One must give security before borrowing or opening a bank account. He must have a deposit to have his check certified. The government banks would ask no more before certifying orders for goods. It would not be necessary to give bonds for every transaction. A bond could be filed for a year or any other fixed period, to cover the largest amount of currency required

at any one time. It might be made low enough, according to the locality, to cover the farmer's load of produce,
or high enough to secure the price of a steamship. If
people would not pledge their goods that would be an
indication that they could exchange them without it,
while no one who had goods to pledge would be allowed
to suffer for lack of money.

All would get money on their goods and refuse to sell?
That objection is puerile. Money is only a means and
a medium of exchange. The business man wants
money for the purpose of supplying his needs. When he
gets the money he will at once go into the market for
that purpose, and where buyers are there also will
sellers be. At least that has always been the experience
of the business world. The very first sale made would
start the ball rolling and matters would go on just as
at present. While each man would be obliged to pay
insurance on his goods and run the risk of holding
them as now, while he would be obliged to redeem
them after a fixed period, he would be just as prompt
as at present to take advantage of the markets and sell.
The amount of perfect security which he could give
above the value of his goods in store would limit the
amount of goods which he could hoard. Each man
would soon reach a point where he would be obliged to
sell.

But, it is said, the establishing of the proposed banking system would increase by thousands the army of
office-holders, and give spoilsmen a power for corruption which nothing could withstand. This need not be
the case. We have already reached a point where good
government loudly demands a radical curtailment of
appointing power and a making of all officials directly
responsible to the people. Elective officers, from the
president down, spend more time in distributing spoils
for the reward of henchmen, or to wrongfully influence
government, than they do in dealing with the real problems of government. The remedy is clearly indicated.
Go back to the principles of local self-government
Make postmasters, attorneys, judges, bankers (under
the proposed system), and all other officers not em-

ployed at the seat of government, elective by the constituencies which they are to serve. Make all department clerks elective and apportioned to congressional districts.

Let us digress a moment and take a glance at the constitutional changes necessary, if the people would retain their hold even on the present governmental powers.

The constitution of the United States, like all schemes of popular government, is founded on the equal civil rights of all men. This implies that all men shall have a share in the government, and that no privileged class shall dictate by the right of privilege what the government shall be. Although the foundation principle can not be denied by advocates of popular government, there were found in preparing the constitution of the United States, insuperable obstacles to its complete realization. A party was found, then as now, who distrusted the people at large and set about placing checks in the constitution which might, under certain conditions, serve to negative the will of the people or set aside a popular verdict. This party seems to have lost sight of the lesson of history: that no ruler has ever governed better than he was compelled to by popular sentiment, and that irresponsible power is invariably used for selfish ends inimical to the welfare of the many. All power not coming from the people is irresponsible, all checks used to negative the will of the people are autocratic. They have proved mischievous as vain. A stream will not rise above its fountain-head, and especially is this true of the stream of popular government.

It was in deference to those who distrusted the people at large that provision was made in the constitution for the electoral college, the legislative election of senators, an appointive, irresponsible, life-tenure judiciary, and an almost unlimited appointive power by elective officers. Experience has shown that these are the very provisions around which cluster nine-tenths of the fraud, trickery, corruption and dissatisfaction that beset popular government.

From the indiscriminate appointive power we have the spoils system, with all the attendant evils of ward politician, voter-corruption and encroachment of executive on legislative power.

From the irresponsible life-tenure judiciary we have shameless obsequiousness of the bench to the money power, the judicial trampling on popular rights, and bench legislation claiming supremacy over all other expressions of governmental power.

To legislative election we owe a plutocratic senate, tainted at its very inception with the stigma of corruption, irresponsive to the wishes or interest of the people, and hedged around with the impotence of obsolete form.

The elective college has practically placed the presidency in the hands of a few machine politicians of New York, Illinois and Ohio.

These things must be changed in the light of the lessons of our experience. The constitution was made for the people, rather than the people for the constitution. The great authors of the document believed that experience would show imperfections and provided a means of correcting them. They recognized the lesson of history that if provision is not made within the constitution for change it will be changed by violence and revolution. Time has shown the imperfections of this wonderful instrument. Will enlightened Americans continue, Chinese-like, to worship error because of its age, or try by the light of experience to make the instrument more perfect, more adapted to the wants of the present?

I do not contend for popular government. In the breast of independent manhood, liberty needs no champion. I speak to those who deny the inherent right of any man or body of men to rule over any other; who look upon a voice in the government as a right and not a privilege; and who consider that civilization without liberty is a hollow farce, an unreal mockery. Not the liberty of licentiousness nor force-bred anarchy, but the equal liberty of just citizens, who would withhold from no man that which they themselves enjoy. Let the man who declares a voice in the government of one's country to be a mere privilege, explain from whence comes the privilege or who has a right to give the privilege, or else go lay his sophistry at the feet of thrones, where reign the crime-polluted *soi-disant* vicars of Providence.

Most of the necessary constitutional changes have already been suggested. They are opposed only by that fossil conservatism which persistently refuses to recognize the fact of national progress. With the race the idols of to-day have been the curiosities of to-morrow. We constantly put away the things of the child, yet a revolutionized civilization blindly clings to the laws of a lost century.

It is not difficult to point out these defects in the constitution. To all thinking men, the vast appointive power vested in executive officers has become a dangerous menace to popular government. It is the foundation of the spoils system, a powerful weapon for the subversion of popular will, a fertile source of dishonesty and corruption. It is probably necessary that the executive be allowed to appoint his cabinet ministers and confidential secretaries, but here his elective power should end. A president would be less than human, or more so, if you wish, if he should refuse to use his patronage power in case of emergency. It has been done and will be done again. Patronage has forced more than one unpopular measure through Congress, and with an able and ambitious executive we might find the legislative branch a mere appendage to the executive. A congressman, unless he be made of exceptional stuff will not oppose a president of his own party, when he knows doing so will conjure up against him subservient enemies in every hamlet in his district. More than that, it will bring government hostility to his constituency, sly discrimination, which will be laid on the shoulders of the congressman, turning into enemies the very persons he has taken trouble to defend. Then attending to the claims of office-seekers is made the chief business of the executive, and multiplied obligations arising from patronage hamper not only the president but every official in the land.

The power may be thrown back upon the people. It would be an insult to every American citizen to say that a time-serving politician, without public interest or personal responsibility, is a better judge of the fitness of a postmaster or other officer at this or that place, than the constituency which that official is to serve. And that is really the issue. Postmasters and other local officers are now virtually appointed by local politicians, who have gained some hold on the administration. The president merely decides as to whose appointee shall have the place. Every postmaster, every customs or revenue officer, every district attorney, district judge or other court official, every mail agent and inspector might readily be elected by the constituency which he serves without at all interfering with the service which he is called upon to direct. This is what must happen if we are to have civil service reform worthy of the name. The principle might be so extended that every department clerk might be elected by the people.

For this purpose the clerical force of the United States might be apportioned among the congressional districts, which might be subdivided into their constituent counties. Three or more candidates might be elected to each vacant position as eligibles, and their final appointment made to depend on an examination intended to test their fitness for the work required. After the examination, the result of which should be made public, the eligible standing the highest should be given the appointment. There should be no discretion left to any one in the matter, but he who stood the best test examination should have the place as a matter of law, the people's vote in electing him an eligible being the only recommendation required. Vacancies between elections might be filled from the other eligibles. These clerks might be elected for long terms, six or eight years, a small percentage going out at each election. In this manner the efficiency of the department forces would be improved and we would have civil service reform in fact as well as in name, and not the farce now foisted on the American people. A more

important consideration than this, we should not have
a class of life-tenure officials, living as sort of beneficiaries of the government and entirely out of sympathy
with work-a-day citizens. These life-officials come to
have interest in but one thing—self, and are ready to
give temporary allegiance to any politician who might
lend them assistance. They know little and care less
about the condition of the country at large. They
make the very poorest and least public-spirited citizens.
And as for the efficiency which comes from long
service, it is not readily apparent. It is notorious that
the want of ambition and disposition to shirk duty developed by long official service, makes the majority of
long-serving clerks less valuable than men of little experience in official life. Let public servants in all ca-
pacities frequently touch elbows with the rank and file
of citizens. It will be found better for all concerned.
Let each one's official record go before his electors once
in a while. It will have a salutary effect. The clerk
would gain, as he would have a fixed term of employment and not be placed on the anxious seat every few
months.

With such an official force as this, multiplicity of
forces would be no objection to the nationalization of
banks, railways or anything else. Besides, it would be
a reform which must be had to save popular government.

The actions of federal judges furnish most flagrant
instances of the mischief of giving official autocratic
power. We have a sort of tacit legal fiction that a
judge, like a king, can do no wrong. Hence we have
made him utterly unaccountable for his actions, for im-
peachment is but a sort of pretentious joke. He is
created by one man with the consent of a body of men,
themselves thoroughly undemocratic, and when once
installed is superior to his social creators. And who
was this paragon of perfection before he ascended the
bench and put on the obsolete insignia of autocratic
authority? A mere lawyer who for money warped and
blocked the laws of his country to further the selfish
or iniquitous schemes of his employers. He was prob-

ably the hired and obsequious servant of some corporation whose profit was wrung from wrongful levies on
the toiling public. He was certainly no more honest,
upright or disinterested than other men, and, no doubt,
often drafted and advocated the passage of laws for the
special purpose of making them dead letters or using
them to his own and his employer's ends when they
came to be enforced.

Is it any wonder, then, that the judge, hedged around
by antiquated forms and illogical precedent, should
continue in his contempt for the rights and interests of
the people and persist in serving the special interests
which have created his fortunes? Is it any wonder that
the plainest intention of the people is negatived by
the judicial dictator and that corporations laugh at
legislative enactments from behind judicial strongholds?
Nobody expects perfection in a lawyer. The ermine does
not transform him. Often the realization of his ambi-
tions depends on how well he serves—some corporation.
There is but one remedy. Make judges of all sorts elective by the people for limited terms and therefore accountable to the people. If the people are capable of
selecting law-makers, they are capable of selecting law-
interpreters. But whether they are capable of so doing
they must judge for themselves. They have a right
to have their verdicts enforced regardless of the opin-
ions of the few as to the wisdom or folly of these verdicts. And in practice courts have not been found to
suffer where judges were made elective rather than appointive, but, on the contrary, laws have been more
intelligently and justly administered. The corrupting
influence of the appointing power on the judiciary of
this country is notorious. To our shame be it said, the
highest tribunal in the land has been bartered as a reward for corrupt help of corporations, or for upholding
the wrongful acts of partisans. But worse than all this
the appointive judiciary has been gradually absorbing
the legislative power, until the country is overwhelmed
with a melange of judge-made law. If democracy is
to survive it must clip the judicial wings, elect judges,
and make them accountable to the people; in a word,
abolish judicial dictatorial power.

The United States Senate, instead of being the lofty, single-minded institution intended by its founders, the institution which was to leaven with wisdom and honesty the crude, passion-laden measures of the popular branch of government, has become the most ignorant, corrupt and inefficient branch of the government. Its actions are more often tainted by suspicion of wrong or dictated by motives of self-aggrandizement or contempt of popular rights, than they are illumined by experience, wisdom or patriotism. Experience has most emphatically proved that those who distrusted the wisdom as electors of the people at large, were entirely mistaken.

We have found in practice that the members of the state legislatures are no more incorruptible, if indeed more wise than the average citizen-voter, and the number is so small that it is feasible for moneyed interests to purchase an election of them. The fundamental mistake is the supposition, negatived by one hundred years of history, that any select body of electors is less corruptible and wiser than the electors at large. The remedy is, then, suggested by history. Make the Senate elective by popular vote and it will cease to be a millionaires' and corporation counsel club, and become a useful branch of the government. One can always attend to his own business better than any one else will, and this is true of the business of the whole people as well as an individual. Make all officers responsible directly to the supreme law-making power, if you would have the will of the citizen carried into effect.

The electoral college has not, for nearly a century, had anything to do in practice with the purpose for which it was established. It was another case of distrusting the verdict of the citizens at large, and speedily grew into the most ridiculous farce ever exploited in the name of popular government. It is now the tool of the New York machine on either side, and gives power to a handful of politicians in that state to dictate a president to sixty-five millions of people. The convention of neither party dares ignore the New York machine in the make-up of the presidential ticket, for, so far as ante-election estimates are concerned, as New York goes so goes the country. The consequence is that the Republican and Democratic machines of the Empire State nominate the tickets for both conventions. Let the people elect a president by a direct, popular vote, and political parties would cease to bid for the support of pivotal states. A million of voters in the Mississippi valley would have as much to say in the election of a president as a million of voters in New York. Political dictators would find a large part of their occupation gone. Presidential candidates would be chosen more with regard to fitness for the great responsibility than to propitiate this or that politician, or this or that special interest in a pivotal state. The will of the people would be carried out as expressed, real majorities would rule, and not crude approximations thereto.

The practical politician is an egoist, so are the interests which he serves. He manipulates politics for gain, for himself and for his masters. His effort to control a political situation is measured by the probable return. It is a money and time investment for a money return, what we call "strictly business." The politician or the agent of special interest supports a candidate for office with the understanding that there shall be a return service. The value of that service depends largely on the length of the term for which the officer is elected. Hence a practical politician will invest more in an officer elected for two years than for one, for four years than for two, for six years than for four, and, in general, for a long term than for a short. Other things being equal, the amount of money invested in election is directly proportional to the term of office to be filled. A certain policy extending over four years will yield greater return than a like policy extending over half that time, and at the end of four years there is a better chance to continue that policy than there is at the end of two. On the other hand the official elected for a long term is more independent of the citizen electors and less responsive to their will. He knows that the money power is with him, as wealth always dreads change. To drive money out of politics and secure a sure and prompt response to popular will, all the terms of all important officials should be shortened by about one-half. No more should be paid for the work of minor officials than is paid for the same class of work by private individuals. There would then be no profit in manipulating politics, and the practical politician is not the sort of man to waste time and money without return.

There would be no upsetting of business by frequent elections, as is often claimed. The very opposite would be the effect. The tenure of party would become so uncertain that every citizen would make it his business to see that the candidates of all parties were trustworthy. The officers who would really represent the voter would have his confidence. Each election would not be a periodic contention of special interests causing a periodic crisis of suspense in business interests. Politics and business would find it necessary and profitable to go on together.

To be sure, with the initiative and referendum, minority representation and other like reforms, short terms would be less imperative, but as a means of avoiding the use of money in politics and preventing periodic business upheavals while awaiting an election crisis, they would be invaluable. Anything tending to break down partisan power and make minorities responsible, would be a boon. The minority would then not seek to make the laws of the majority as bad as possible in order to cause a revulsion of feeling and a return to power of their party.

This is in the nature of a digression, but still germane to the principal subject. These reforms must be made sooner or later, whether we nationalize the banking system or no, and if they are made, the only valid objection to extending the powers of government would vanish. Where the government is really and truly the people, there is nothing which the people wish to do that they cannot legitimately do without violating a single principle of individual rights. It is only to governments extraneous to the people that the criticism of the danger of too large an assumption of the power of the government holds valid.

To return to our principal theme, the proposed warehouse receipts might be made in amounts large enough in no way to obstruct sales and still give every one the advantage of the accommodations of the bank. The warehouses need not be different from the bonded warehouses of to-day; the duties of inspectors, gaugers and weighers would be less difficult than the duties of our present custom and revenue officers.

The receipts, as stated, might be issued in duplicate, so that the bank may retain the original, while the duplicate could be made negotiable, thus accelerating sales. The duplicate should, however, be made simply an evidence of ownership of the original, and the original alone should be the instrument on which the goods could be drawn.

Security other than the goods for sale would be required of each patron to the full amount of his bank credit, to secure the bank against possible fraud in weighing or inspection of the goods, and to enforce the requirement that the goods should be sold by their owner, thus relieving the bank of all unnecessary trouble or responsibility. When the goods were sold the credit would be canceled to that extent, and a volume of money corresponding to the goods sold would be withdrawn from circulation.

Bonds and stocks and evidences of credit may very properly be taken as security, but no money should be issued on them, for the simple reason that they are not wealth. That is the fatal weakness of all money systems so far devised. The volume of the debts of the United States as a nation has no necessary relation to the current business of the country and hence to the volume of currency required. The volume of public debt or even private securities has no more relation to the exchanges of the country than they have to the necessary transportation facilities. A money founded on them is entirely without a criterion as to volume. It may be twice too large in amount and may not be half large enough. And besides, a money founded on wealth could not be based on evidences of indebtedness. It would be paper founded on paper, not wealth in any sense. In the sale of a railway or a piece of real estate, to be sure, the instrument of title could be taken as evidence of ownership, and be treated as such.

In practice it would not be found difficult to make goods in transit the basis of bank credit, so that commodities intended for a distant market may be forwarded to that market and at the same time serve as the basis of a money or credit issue. The intimate communication of the banks with one another and the ease with which the carrier may be made to serve the purpose of the bonded warehouse, would make such credits especially easy to arrange. The details need not be given.

Of course the banking system would not be a money-making concern, and a small percentage on business transacted would give a revenue sufficient to pay rents, clerk hire, and cost of printing books and notes.

There could be no objection to making deposits of currency the basis of bank credit.

It would be introducing no new principle to give the government through its banks full control of the issuing of all money, and the making of what it chose legal-tender. The so-called precious metals must cease to exist as legal money before any reformed currency would be perfectly successful. Otherwise private contracts wrung by the controllers of gold would force the currency back to a metallic basis. Of course gold and silver, like other wealth on the market, should be allowed to become the basis of a money issue, but merely at their market price.

For the convenience of foreign trade as well as for its own protection, the government would continue to weigh, test and stamp with assay mark, gold and silver bullion. This would make the bullion as readily negotiable in foreign trade as our present coins.

It would probably be necessary at first to give the government a legal monopoly of the banking, as it has of the postoffice business, but this necessity would, in time, pass away, as there would be no occupation for private bankers.

CHAPTER XXXI.

Many men prefer doubting to understanding.

WITH a perfect national banking system such as proposed, commercial loans would be entirely done away with and but little interest, comparatively speaking, could be collected. But currency is generalized wealth, or rather an order for any sort of wealth on demand, and it is not subject to the universal law of wealth, the law of decay and final destruction. It is not bulky, not difficult to protect. Hence where the demand may be made at any time for an unlimited period, and the face value collected, the currency would be more advantageous for hoarding than any form of wealth, and the receivers of surplus incomes would withdraw it from circulation and hoard it for future use. It would represent a demand not made and hence that amount of goods would remain a glut on the market. Demand and supply would be so far unbalanced. The hoarders of currency would shift the burden of wealth deterioration on some one else. They would finally hold enough currency to derange trade somewhat and make it necessary for others to borrow at interest, even in commercial transactions. They would lend at interest to new enterprises It is to obviate this that I suggest the specific laws against interest-taking.

It has become fashionable to assert that prohibitive statutes or statutes limiting the amount of interest taken are not only ineffective but mischievous. Even if this were true it would have nothing to do with the proposed prohibitive statute. The trouble with usury laws so far, is that they have been mere acts of expediency founded on no principle whatever. No moral turpitude attaches to the taking of interest to any amount. If it is not wrong to take six per cent interest when one can get that much and no more, it cannot be very wrong to take ten per cent interest when one can get that amount. If taking ten per cent interest is not wrong in California it is not wrong in New York. The laws recognize interest-taking as right and fix a sort of barrier lest too much right should make a wrong. It would be as sensible to regulate the amount of beating which one citizen might give another without making himself amenable to the law.

Again the legislators who pass these laws, the lawyers who draft them and the judges who are called upon to enforce them, believe in the divine right of every one to take all of another's goods that he can get, and evade the law. In short, the laws are wrong in conception, inadequate in provision, and untrustworthy in enforcement.

Yet with all of their imperfections, these laws, when honestly administered with the intention of suppressing what was thought to be excessive interest, had a very marked effect. The Queen Anne statute of usury was rarely evaded, and succeeded in cutting down the ordinary rate of interest to five per cent. It was repealed on sentimental principles of what was posited as freedom of contract rather than from any inefficiency or hardship due to the law itself; and its repeal worked a hardship on all except those with the very best credit. Of this law Lord Mansfield said, "Where the real truth is a loan of money, the wit of man cannot find shift to evade the statute." In controverting the opinion of Dr. Smith, that no law can reduce the rate of interest below the ordinary market rate, Mr. Bentham in his "Defense of Usury" says: "As to the general proposition, if so it be, so much, according to me, the better, but I must confess I do not see why this should be the case. The evasion of the French law might be due to a defect in penning that particular law, or in the provisions made for carrying it into execution. In either case it affords no support to the general proposition. For the position to be true the case must be that every law would still be broken, even after every means of what can properly be called evasion has been removed. For destroying the law's efficacy altogether, I know of nothing that could serve, but a resolution on the part of all persons any way privy, not to inform. In England, as far as I can trust my judgment and the general recollection of the import of the laws relative to this matter, I should not suppose that the above proposition would prove true."

As a matter of fact, Bentham was right. Real estate mortgage loans never, while the statute was in operation, drew more than five per cent interest, while it was proved before the committee of the Commons who investigated the matter in 1818 and the committee of the Lords who investigated it in 1841 that the market rate of interest had frequently been more than five per cent during that period. Never did bond or note for the century and a quarter during which that law held sway, bear more than five per cent interest. Mr. Byles, in his work on usury, quotes Mr. Gurney, an eminent English bill broker of the time, as saying: "I consider the evasion of the usury laws to be partial. I am of the opinion that the lender of money was obliged to be satisfied with five per cent and that the borrower has got it at a more reasonable rate in consequence of the law." The same author quotes Mr. Kemphle, an extensive produce broker, in business for more than thirty years, as saying, "I am not aware of any practice among merchants to avoid the usury laws, or attempt to avoid them." The testimony of Mr. Gibbs, the great annuity dealer, is in the same line, and is corroborated strongly by Mr. Maynard.

I cannot go into the history of legislation against usury at length, but I have quoted far enough to show that where laws against excessive interest were made to be enforced, and the people concerned wanted them enforced, they were enforced, at least as thoroughly as any other law. We would not think of repealing laws against stealing or fraud because such things are still practiced, yet if the argument of ineffectiveness is good in one case it is in the other.

Make a usury law founded on the theory that interest-taking is wrong in itself—a crime. Draw it so that it will be difficult to evade, and support it with a public sentiment as strong as that against stealing, and we would have no trouble in having the law enforced. That is what I propose. There are people who have such a mistaken idea of honor that they will refuse to bring to justice the sharper who robs them. There are others who would suffer rather than have it known that they are dealing with usurers, but these are a small percentage. If half of the forfeited debts should go to the informer and half to the state, self-interest would largely provide for the enforcement of the law against those who loaned surreptitiously.

The provision making loans between private parties uncollectible unless made through the government bank would make the evasion of the law by subterfuges entirely impossible. The bank officers would deliver the instruments of security to one and the money to the other, and could easily see that there were no deductions. This is no more than we now compel in the transfer of real estate. It has to be done in a certain fixed way to make it valid. A watch over the contract of loan is just as important.

Here again we meet with the objection of Bentham and others that the state has no right to interfere in contracts between individuals and that its interference always works evil. Our government is constituted largely for the purpose of protecting the weak from the aggression and wrong of the stronger, and anything which will do this is certainly within the province of government. Our sanitary laws, our coinage laws, our laws against adulteration of foods, our statutes against fraud, our statutes against gambling, our regulation of rates for carriers, and thousands of other laws are made to interfere in individual contracts in behalf of the less wise, less cunning, or the weaker party. If government is not going to give greater security to citizens in their rights, then government has no excuse for being. And so far as being an interference with individual rights, it is no more an interference with the rights of citizens to prevent them being defrauded by usury than it is to prevent them from being robbed by force. On the other hand, no one has a right to wrong another. And in a government of the people, who is the judge? Who is to set limits as to what the people choose to do in their corporate rather than in their individual capacity? Certainly no one but the people themselves. If they think their interest can be better protected by preventing one man from taking advantage of another, they have certainly the right thus to protect them. No man has a right to become the object of a wrong, and it is absurd to say that you invade one's rights when you protect him from wrong. And the law could not even be passed without the support of popular sentiment. Anything which will better the relations of man to man is the province of government.

It is scarcely necessary to go further in answering the apologists for usury. Bastiat is the only one who has gone at all into the philosophy of the subject. He has been effectually disposed of. Bentham goes no further than to show the inconsistency of the average laws against usury. He says that as the percentage of interest charged must depend upon circumstances, where it is right under one set of circumstances to charge five per cent it is equally just under other circumstances to charge ten. Where we make an inflexible rule we work injustice. If six per cent is not wrong, ten per cent can not be very wrong. Such argument has no force with one who denies the right of any borrower to take any interest whatever and considers one per cent or ten an unwarranted extortion and moral wrong.

Bentham thinks that laws against usury are further

inconsistent in preventing interest on money and allowing hire for the use of all other sorts of wealth—rents
for houses, etc. There is truth in that position. It is
inconsistent, but the proposed scheme does not embrace
that inconsistency. It denies the right to take increase
for any sort of wealth, "for anything that is lent upon
usury."

Then Bentham says that a certain class of borrowers
in desperate circumstances will, under laws against
usury, be obliged to go to the wall at once and not be
able to prolong their business career until the usurer
has all of their property and the rest of their creditors
are left without recourse. This is the real substance
of his argument about the impossibility of persons borrowing with little or no security. There is no hardship
in that. Without usury laws such persons will be
charged rates which no business can stand longer than
a few months, and their property will go to the money-
lender. It would be better for all concerned if they
could not borrow at all. Then their property would
be saved for their legitimate creditors. Under the
proposed system, no one with security to give need want
for money.

If money may be hoarded with impunity it will be
hoarded, for reasons given above. The only way to
prevent such an outcome is to make a currency the holding of which is no more profitable than the wealth
which it represents. Nothing but a time-limit currency
could be made to deteriorate with holding beyond a certain limit without impairing the use of that currency as
a circulating medium. With a time-limit currency, the
whole volume issued during a corresponding period must
go back to the bank of issue before the time-limit ex-
pires. Those who simply used their money for hoarding could then be taxed an amount sufficient to make
up for the deterioration of the wealth represented by
the money they held. They would then use their currency or lend it without interest, and save the amount
of the deterioration tax.

It would be entirely feasible to stamp notes with the
date of expiration so that any one able to read currency

could tell the date. The money obtained in the regular course of business by persons holding bank credit on goods would be redeemed as a matter of course in the currency of current issue. The currency should be legal-tender for debts up to the time of its becoming non-current and hence it could not be discriminated against. The low limit which would be set to the amount redeemable free of charge for any one person, coupled with the requirement that all currency presented for redemption should be left at the bank until the period of redemption had expired, would make it impossible for the holders of large amounts of currency for hoarding, to have their holdings redeemed piecemeal. On the other hand, it would relieve from loss wage-earners and small shop-keepers who did not carry bank credit and received the currency in the course of business, and thus prevent the currency from being discriminated against in trade as well as in paying fixed debts. Experience would teach the proper limit for the free redemption of currency. It may be fifty and it may be two hundred dollars. In practice nearly the whole of any issue would be taken up before it became non-current.

This provision is, of course, for the purpose of preventing any one from hoarding currency, and compelling the holders of surplus to lend it without interest.

A charge of percentage on deposits would be for the same purpose. It would make the holding of currency as expensive as the holding of real wealth. Where one held wealth as the basis of his bank deposits or credit, it would be manifestly unjust to charge him a percentage on his credit as well as compel him to take care of his wealth. This provision would place the holders of money on the same footing as the holders of wealth, nothing more.

After the days for redemption at par had expired, then the non-current notes hoarded would gradually lose their value, just as the corresponding wealth would deteriorate until they would become worthless in time, just as wealth does by holding out of use.

Adam Smith did not go into the philosophy of inter-

est-taking. He seemed to think that competition between lenders would be sufficient in older countries to
keep down the rates of interest to a reasonable figure.
He seemed to forget that most of the competition is on
the other side, among the borrowers, and that the rule
among bankers and financiers generally is close and
effective combination. His arguments against limiting
contracts are the stock article already noticed.

But even if interest were lowered to the legal rate of
most countries it must still be an extortion. If wealth
does not make wealth, as I believe I have proved it
does not, then the taking of any increase is unreasona-
ble, and we have never come to the point where money
was loaned for nothing. But say Smith and many
others, the borrower makes a profit on this borrowed
wealth, even after paying interest. So much the worse.
Where does his profit, including the interest, come
from? The wealth used to pay a profit did not make
itself, it must have been earned by some one. If not
by the borrower, then by some one else to whom it would
rightfully belong. The very term "profit" implies unjust
advantage. It is gratuitous gain for which no return
is given. Before we reckon net profits we deduct all
legitimate expenses and then we have left a shave, an
advantage wrung by the cunning of some one from the
labor of another. All profit is really interest, but as it
cannot be attacked directly we must try to reach it
through the medium of interest on money loans. The
currency system is the key.

It is very generally but very gratuitously asserted
that we shall finally reach a point where abundance of
capital will make interest merely nominal. Let us see.
According to the Old Testament, usury taken by usurers among the tribes of Israel often amounted to not
more than one per cent. We have no record of interest
being so low afterward.

Interest in Rome during the empire was down to four
per cent. The average interest in America to-day on
perfect security for moderate periods is six and one-
half per cent. Slow progress this toward no interest.
There are instances of interest being so low as three

per cent in England one hundred and fifty years ago. It was even lower in Holland before that time. At the time of the passage of the Queen Anne usury statute, five per cent was considered the largest amount which should be taken in interest. In not one of these countries can cheaper loans be had to-day except on very long time. Government loans are exceptional. They carry with them a sort of assurance that in rapidly changing conditions the man of wealth will not be disturbed for at least a long period. Where one is so enormously wealthy that a small percentage will give him an all-sufficient income without any trouble whatever, he naturally prefers such an investment to a better paying one without the assurance and with the element of personal exertion. Then, government bonds not being subject to taxation, the net interest charge on money loaned on them is not so very different from that loaned on other property.

We are constantly told that interest in this country is gradually falling. As a matter of fact, it has increased pretty steadily since 1880, if not since 1873. According to the mortgage researches of Carrol D. Wright, the nominal interest on farm loans has decreased one-tenth of one per cent since 1880. The nominal interest on lot mortgage loans has decreased a little more than three-tenths of one per cent in the same period. According to the calculations of experts, gold has increased in value between fifteen and twenty per cent during that time. At the lower figure the real interest charged for farm and lot mortgage loans in 1890 would exceed the real interest paid on such loans in 1880 by about one per cent. For all interest is contracted for and paid on a gold basis. This will put the average interest on land mortgage loans up to eight and one-half per cent, a formidable figure.*

If we apply the same test to government loans, we shall find that our advantage is not nearly so marked

*As the average period of a loan is about five years instead of twenty, the actual increase of real interest rate due to the appreciation of gold is but one-fourth that given in the text, but this is sufficient to more than counteract apparent reduction in rate. Besides, when the principal comes to be paid, from one and one-half to two per cent per year is added by reason of money-unit appreciation.

as one might imagine. Economist McCulloch says, in a note to Adam Smith's "Wealth of Nations,"that it is doubtful if a lower rate of real interest is paid in England to-day than was in barbarous times. The same is true everywhere as a general proposition. There are so many laborers on the verge of starvation and their competition for the use of wealth so fierce that it is impossible, under present conditions, to have interest materially fall.

Others, again, learnedly divide interest into risk-premium or insurance and interest proper and maintain that while interest proper is everywhere, in this country, practically the same, that risk-premium gets the rates distorted entirely out of proportion. It is a very good explanation except that it is contrary to facts. As regards a certainty of having the money of the borrower refunded, real estate mortgages are the most perfect sort of private security in general use. Real estate is not mortgaged for more than one-third to one-half of its market value; in fact the figures of the government bureau of statistics put the percentage much lower. The entire cost of making the loan is borne by the borrower, including all brokerage and commissions. Where, in the name of common sense, is the risk? Yet when money could be had by the government and some favored corporations for three and one-half per cent, six to nine per cent was being charged on mortgage loans. In modern transactions of the more legitimate sort, risk has nothing to do with the rate of interest. The security is believed to be perfect and is perfect usually before the loan is made, except in the case of five per cent a month sharks entirely beyond consideration. And even these seldom lose.

There are two or three reasons why interest rates vary. Our financial system and financial policy tend to mass all surplus wealth in a very few commercial centers of the East. Here it is remote from a mass of borrowers on mortgage security, its possessors know nothing of and care less for the condition of the users of the surplus and allow their wealth to seek more or less remote markets only on the condition of offers of

large gain.   For the mass of borrowers there is absolutely no competition among lenders.   They cannot
afford to go into the money markets of the East and
bid, and are at the mercy of an agent who dictates rates.
On the other hand, the United States government can
go into the markets of the world and receive the benefit
of world-wide competition.   The same is true in a less
degree of all large organizations.

Again, the extremely long time which government
and like loans run, give them a permanence which is
everything with the great ease-loving capitalist. He is
sure of a certain amount for a large portion of his lifetime, while, if he had to renew every five years, who
knows what changes might take place in the meantime?
Then at present government bonds are so convenient as
a basis for money-making banking enterprises, and are
exempt from taxation.

Any observer may verify these statements for himself.   During the financial stringency of 1893 a loan
was sought on a piece of land inside of Fourteenth
Street and between Mill Creek Valley and Olive Street
in the city of St. Louis.   As is well known by all ac-
quainted with the city, and as may be verified by any
one with access to a St. Louis map, this land was but
just outside the most active business portion of the city
and there was no possibility of its deterioration in
value.   It was worth then, would sell at auction for,
twice the value of the loan asked.   Yet the loan was
refused at eight per cent   Call loans were made in
New York that week at a figure scarcely above the average.   Was it the risk in the premises which put up
the rate of the St. Louis loan?   Not at all.   It was the
manipulation of the money market, and the massing of
the bulk of circulation away from the localities where
it was needed.   This is a typical, not an isolated case,
and the city named was probably less distressed in that
way than any other city in America during the recent
panic.   The risk explanation has not a leg to stand
on.   It is a mere trick of usurious ingenuity, unthink-
ingly adopted by economic writers.   It might have
some force at times of governmental insecurity or in

early periods of violence.   It certainly has not at present.   The man who loans at ten per cent on the Minnesota farm is as sure of the return of his capital and interest as the man who invests in government bonds at three per cent.

CHAPTER XXXII.

If Reginald gets something for nothing, Jonathan must give something for nothing.

To the reader who has followed me through these pages, it is by this time clear that interest-taking is wrong in itself. It is also clear that interest-taking is at the bottom of much of the industrial wrong and failure experienced in the world. I have pointed out a means of avoiding that wrong. The only help for the oppressed industrial workers is to adopt this or some equally effective means of destroying interest-taking and placing the world on a just and sound industrial basis.

We have seen that with the present effectiveness of machinery, there is but a limited product to distribute between the toiler, landlord, and the capitalist. If the toiler would get more the others must get less. The others do not earn what they get. The laborer has earned a right to all that is left of the gross product after keeping up capital. If he would increase his wages he must take what he is entitled to. There is no magic by which larger wages, as well as larger profits, can be paid out of a fixed product, or a product increasing only in proportion to the march of civilization and population. The laborer must realize this and strike at the root. All unearned incomes are founded on rent and interest-taking. Any unearned income makes the wages of the laborer less, for it is getting something for nothing. It is at unearned incomes, then, that the laborer must strike and the place to strike at them is at

224

their foundation. He must destroy rent and interest-taking by private individuals if he would better his own condition. As for earned incomes, they can never be a harm to any one who earns what he gets, for they imply that the receiver gives full value in return. There are no incomes earned by wealth. All that is earned must be earned by human labor of hand or brain. The cutting off of all incomes from other sources must take place before those who labor will get all that they produce, and this can be done only through the destruction of interest-taking.

In a just industrial organization there can be no such thing as capital profits. Profits mean something for nothing. They mean that after one is paid for all outlay and all services, he is paid something extra, for nothing, a sort of bonus, an advantage over those with whom he is dealing. It would be impossible that all should have such an advantage. One must get the advantage from those with whom he deals, and if he gets something for nothing, some one else must give up something for nothing. A rule of economics which cannot be generally applied is defective. A rule which posits gain for one · at the expense of others, is unjust. The laborer, from the nature of things, gets no profit. He gets but his hire, just what he works for, and that is but a portion of what his labor produces. He never can get profits, for he will never be given more than his labor produces. Non-laborers are capitalists or landlords, and it is they who must get the profits. The laborer produces all wealth, for wealth does not make itself, and hence must pay to the landlord and capitalist their profits. Profits are founded on interest-taking. They are founded on the assumption that wealth is in itself productive, and for that reason goes to the possessors of wealth alone. They can be destroyed only by destroying interest-taking and making all persons workers. Present laborers must have help to bear their burdens.

This is the salvation of the laborer. It is the task which he must perform if he would save his industrial, and finally his political independence. Strikes are as extravagant as foolish. Let him learn just what he wants. Let him then make a specific programme and insist at the polls that it be carried out. After the political changes which I have suggested, the government will be his. Let him use it. It is his only hope. Concentrate on one principle at a time. Let every one who labors support it, and the laborer will soon have his rights.

It is said that well-fed laborers, according to the Malthusian doctrine, will increase so fast that their progeny will soon overrun the earth and the over-crowding will make them as poor as ever—that population will always outrun subsistence. That is a doctrine full of unction to the top dog, the apologist for the system that exists, but it is contrary to experience. Observed facts teach us that with the human family, the best fed, most prosperous and intelligent, increase more slowly, and the best check to population is to make the whole race prosperous and intelligent. It is needless to cite instances. It is a matter of common observation. Let the toiler destroy interest-taking and organize industry on a just basis, and he can afford to be indifferent to consequences. Right never produces wrong.

CHAPTER XXXIII.

There is room for him and for me.

HERE, then, is outlined a thoroughly scientific and feasible currency system. A system which will respond to supply and demand as readily as the mercury in the thermometer does to changes in temperature. A currency which will be perfectly stable as to unit of value, which cannot contract so as to produce scarcity, which cannot expand so as to produce over-supply. There are no untried principles introduced. The proposed currency is founded on value, and the volume is regulated by business needs. It will be neither for gold-bug nor silver-bug nor paper-bug. It will be a scientific currency, recognizing the fact that we have gone beyond the stage of barter and can stamp orders for wealth on articles of no intrinsic value.

It may be put into effect immediately with less expense and less friction than would be encountered in changing tariff or revenue laws. The value of the precious metals which it would take from sequestration would bear the expense of the new system twice over. Banks could be allowed to close up their business gradually and there need be no panic except of their own making, which would recoil on their own heads.

We would be relieved of the incubus of gold-worship and our civilization and prosperity would be allowed to expand beyond the gilded fetters which at present bind them. The system could be put in perfect working order within a year. Within five years the country would have adapted itself to the new conditions and all would be peace and prosperity.

Better than all this, usury would be dealt its death-blow. Each man would have what he earned, no more. The laborer would have the share which now goes to the usurer and ere long would add to that the share which goes to the landlord. All mankind would be toilers with common rights, common interests and common sympathies. Prevent the tiger from preying and he will become as docile as a kitten. Give both security and the lion and the lamb will lie down together. It will not make a paradise. The lovers of power at the expense of fellow-beings, the men who conqueror-like mount the ramparts of fame on the life-less bodies of their fallen kinsmen, will want none of it. For the man who gives a return for what he gets it is a start toward a nobler goal than he has ever yet aimed for.

Vested interests will be proclaimed. It would be a contradiction of terms to say that one had a right to wrong a fellow-man. Yet this is what vested interests imply. If the taker of interest is wronging his fellow-man the fellow-man has a right to have that wrong abated, and this is the only vested right which a just civilization can recognize. It does not make it better to say that the government has from time immemorial allowed the ancestors of the usurer to wrong the ancestors of the laborer. That is all the greater reason why the wrong should cease and the conscious or unconscious wrong-doer be satisfied to let bygones be bygones. Even if in our imperfect courts of law wealth has been wrong-fully converted, restitution will be compelled of the innocent beneficiary. We cannot cavil at the justice of a similar proceeding in this case. But restitution is not asked. Let bygones be bygones. No one but a fool or a knave would talk of the division of property. All that is asked, is that henceforward every one be allowed to keep his own and no more. The effect would be magical.

But now comes our reform friend riding on his hobby, an excellent nag, to be sure, but not able to bear all burdens. "My reform will make yours unnecessary. Follow me!" Mistaken enthusiast, no one reform can make the world what it should be. There is more than room for all. Every reform founded on truth is needed. Every wheel of the social car must run in the groove of natural law if we would avoid soul-destroying friction and frequent wreck. Let us not be narrow enough to place one little truth before our eyes so closely as to shut out the sublime possibilities of God's universe.

This sermon is particularly intended for the single-tax advocates who seem to pooh-pooh aught else than their favorite theory, and to maintain that after that reform is accomplished nothing more will remain to be done. As I take it, the doctrine is not preached by the leader so much as by the followers. As I hope I have already made evident, I have not the slightest intention of opposing land reform. It is both necessary and fundamental and perhaps in strict philosophy should come first, but it is no more fundamental nor important than the proposed reform. In the light of both the past and the present, I would judge that the usurer has done and is doing more harm than the landlord. It was the usurer who, through usury, came to possess the land, rather than the landlord who became the usurer. How land reform could reform currency or destroy interest-taking, is not at all evident. Its greatest apostle thinks that it would increase interest and I agree with him.

As is stated above, land reform in itself is a consummation devoutly to be wished. A single-tax on land, if it is possible, is the most direct, certain and economical tax and therefore the best. It is the tax which would prove the death-warrant of the public squanderer and petty tyrant. But can a single land tax be so levied as to take land from the grasp of the monopolist and speculator and deliver it over to the toiler? Will it nationalize land? Here is the rub. Here is where the remedy must fail of the ideal completeness which is claimed for it by its enthusiastic supporters.

The conclusions arrived at by single-taxers are much better than the reasons which they give for those con-

clusions. They are entirely right in the idea that the
land should be practically, if not actually the common
property of all. They are entirely wrong in the assumption that land values are not the product of the toil of
hand and brain, and that nature has a greater share
than man in the production of wealth.   They are still
wrong when they assume that it is because the laborer
has no share in the production of land values, that he
is not entitled to remuneration for allowing these values to be used.

If the contributions of nature and of man himself to
the needs of the race are to be compared at all, we must
set what nature offers without any aid from man, over
against what man adapts from the stores of nature.
We must compare what nature does *through* man with
what nature does *outside* of him.   Otherwise our comparison can mean nothing, for man is a part of nature
and subject to its laws.   All that he has, all that he is
or ever can be, comes to him from nature.   Looking
on the matter in this light, nature does all either
through man or without him.   But if man is to be an
element in the calculation at all, we cannot look upon
it in this sense.   We must give man credit of doing all
that is done *through* him, if we are to give him any
credit at all.   There is no other basis of comparison.
In that sense man produces all wealth.   Production is
adaptation.   Nature furnishes the material to adapt,
but this material has no value until adapted.   It is the
adaptation, and the adaptation only which gives the
free gifts of nature (exchange) value; and exchange
value is the only value with which political economy
has to deal.   Nothing but human labor can adapt anything to satisfy the wants of man.   Nothing else can
produce wealth.   It is the act of plucking the bread-
fruit and eating it which satisfies primitive man's hun-
ger.   It is the labor of the herdman, the spinner, the
weaver, etc., which turns the wool into a coat. Nature
invariably resists this adaptation, sometimes more and
sometimes less, so that it takes an exertion called labor
to accomplish it.   In fact the very thing which dis-
tinguishes wealth from non-wealth, is the circumstance

that it was produced by man and not supplied freely by nature.

If this be true, and it is incontrovertible, unimproved land has no value and never can have. (It has neither exchange nor rental value.) Any land which has value must have been improved. And this is strictly true in fact. All land commercially connected must be considered as a whole. The surveyor's lines no more divide land, economically speaking, than the equator or the tropic of Cancer. Any improvement in any part improves the whole and it is improvement on some part which gives value to the whole. Improvements in New York City give value to land in Missouri. The value may be almost infinitesimal, but it is still given. Thus the multitudinous improvements and adaptations on earth incident to human civilization, and these alone, give value to land. This civilization and the resulting value which it apparently gives to land is as much the product of toiler's hand and brain as is any dwelling, any machine, any piece of work in existence. It is contributed to by the toilers in the exact proportion in which they produce wealth, and is as much theirs as is anything which they produce. The wealth or advantages produced in common by the exertion of all attaches itself to the land. A greater amount attaches itself to the land nearest the centers of greatest improvement and that offering, for other reasons, the least resistance to adaptation into wealth. Other things being equal, land in New York City is more valuable than land in Hoboken, because it is nearer the center of greatest improvement. For farming purposes, land in Jersey may not be as valuable as land in Pennsylvania, for its proximity to market may be offset by the greater resistance which it offers in raising a crop. What can an isolated trapper, while he remains isolated, get for a sealskin? Nothing. It is useful to him. So is the land to the isolated settler, but neither has exchange value. Both increase in exchange value, both increase in utility, as the isolated settler or trapper approaches the community. The skin can satisfy more wants in New York than in Alaska, the land more in

Jersey than in the wilds of Africa. They both have greater value in every sense. That value is given in both cases by the civilization built up by the labor of the community. Value is a product of society.

The values contributed by the individual to the common reservoir of civilization attach to land; the values created by individual exertion for individual use have land attached to them. The former can be enjoyed by access to land. The latter must come actually into the possession of consumers. By assuming exclusive possession of what is produced by individuals in excess of what is needed to supply their immediate wants, capitalists are enabled to shut out laborers from the use of that portion of their patrimony which has crystallized into machinery and improved processes of production. By assuming exclusive possession of the land, landlords are enabled to shut out from them the use of that portion of their patrimony which has collected into the common reservoir of civilization, and can be enjoyed only in connection with land. But they do more than this: they shut them away from the land itself, which is the free gift of nature to all. Land values and all other values are created in precisely the same way. They both belong to the producers, but only in usufruct. The use of wealth other than land necessarily carries with it possession, and if the producer wishes to use it he should certainly have possession of it. He has a better right than any one or every one else. The use of land values carries with it, not the possession of the values themselves, but the possession of some of the land to which they attach. They are always ready for him who wishes to use them and has access to land. As the values attached to land are no more productive than the values attached to any other form of wealth, no matter how large a share one has had in the production of these values, he is entitled to no remuneration for having these values used by some one else; provided always that he is allowed to use them to the full extent to which he is capable of using them in production

There are pieces of land to which attach a greater amount of this value which is the common property of

all, and pieces of land to which attach a less. That
is, in the adaptation of nature's gifts, there are lines
of greater and lines of lesser resistance, depending on
the locality of the process of adaptation. The advantage of the locality of least resistance over greater, is
the rental value of land. The discovery and utilization
of these advantages, like the invention of machinery or
the adaptation of any other of nature's forces, is due
to the cunning of man's hand and brain. But it is due
to no one man and no one man should receive a special
advantage therefrom.

If the multitudinous improvements and adaptations
produced by toiler's hands and brain, and incident to
civilization, have produced so-called land-values, then
population has not. Increased population is but an·
effect of the same cause which produces increased land
values. The erroneous idea that demand determines
value, is at the bottom of the error. Demand is nothing
more than a regulator, like a governor on a steam engine. The labor required to produce a thing determines
its value.

The land-values are produced by toilers just as truly
as are the values of anything else. They should not be
parceled out among toilers, because it is impracticable,
as well as unnecessary to their full legitimate enjoyment. Rent should be left in possession of the community, not because it is not produced by the individuals
of the community, but because it can best be distributed
among those who produced it, by using it for the common benefit of all. It is merely ridiculous to hold that
men have any right other than in usufruct to other
property, and to deny their right to land values. Their
rights to both are exactly the same and spring from the
same source. They have no greater right to one than
the other. They have a right to remuneration for the
use of neither, for neither is of itself productive.
Single-taxers cannot deny these principles and still retain a basis for their theory. They cannot condemn the
landlord and shield the usurer even in theory. The
principles which condemn interest must be invoked to
condemn rent-taking by private individuals none others

will answer. The problem before single-taxers is to make land values truly the common property of all. Will the single tax accomplish this?

We must take people as they are. The passage of a single-tax law will make the average citizen little better than he now is. It may, to be sure, be an evidence that he is advancing in enlightenment. After its passage we have merely a law to be interpreted and administered by men. To say that the result will be perfection is like failing to allow for friction in estimating the working power of a machine

To make the remedy complete the whole rental value of the land must be taken in taxes, no more, no less. If more is taken it will ruin the business conducted on the land. If less, it will leave a margin of profits to the landlord and speculator. The assessor must, then, estimate very closely what the actual rental value of the land is. Can he do this? Will he do this? He does not do it now, he does not begin to do it. In some instances he gets eighty per cent of the value of the land and in some not twenty. He may even assess a piece now and then for more than its actual value. Now he has a perfect guide to go by. The exchange value and the current interest on investments give him data which will lead to most accurate judgments of the rental value of the land. The selling price of the land is a perfect index of its assessable value. He fails now through incompetence or dishonesty. These elements must be allowed for in the assessor of the future. And what tangible criterion will he have on which to base a judgment of taxable valuation. None at all. If the remedy is of any avail it will destroy the value (exchange) of land entirely. One can sell land itself for practically nothing. He will have simply the owner's statement as to what the land is worth to him, on which to base the assessment. He may assess a piece of vacant land as highly as a piece of improved land beside it, but as for assessing the improved land he will be entirely at sea. In the rental of improved lands the land and improvements are hired out together. The assessor might by an inquisitorial process find out the income of each piece of improved real estate. Then he must find out how much of that income is due to land, how much to improvements. He can arrive at a conclusion only by closely estimating the value of improvements and deducting the amount of the income due to them from the whole income. He must take the owner's statement for this, for we have yet found no better way It would involve all of the difficulties which render the levy of personal taxes at present so annoying and unjust.

We might say that the assessor might judge by what others paid. This begs the question, for the others must first be dealt with. We might give the land to the highest bidder. This would not help the matter. If a bidder take improved land he must pay for the improvements at the owner's price or the state must compel the owner to sell at an arbitrary valuation. The former would defeat the law. The latter would introduce the principle of taking private property for private use, a very dangerous innovation.

It is ignoring facts to say that the rate of rent is not now estimated on the basis of exchange valuation. Every man counts his land in dollars, as he does his capital, and estimates that it should bring a certain return. If all were perfectly willing to tell all they know, the assessment might not be so difficult, but they are not. It is quite high-sounding as an oratorical flourish to say that men, like hogs, must be led to reform by the prospect of gain. This is what the single-taxers say they offer. Gain for whom? If it is gain for the many, then the answer is that the many are often blind to their own interest, and the best of laws may fail through ignorance. All just laws are for the benefit of the many, but if all just laws were carried out in their letter and spirit there would to-day be no place for the reformer. Then the many are not they with whom the single-taxer has to deal. The many are not enriching themselves on land rents at the expense of the few. If the few is meant, then the proposition is false. The single-tax or any other just law is not for the personal interest of the few who fatten at the expense of their brothers, and can not appeal to them

on that score. At least it is not for their pecuniary
interest, which they consider paramount to all others.
The single-tax law is for the especial purpose of cir-
cumventing those who thrive at the expense of their
fellow-men, and obliging them to relinquish their unjust advantages. On that score it will be necessarily
fought to the bitter end by all beneficiaries of the present system. The same opposition, no more, no less, is
encountered by all just laws.

Then improvements are at present made only on very
long ground leases, fixed as to rate for definite periods.
Would private parties or corporations improve ground
in the possession of which they were secure for but a
year? Could security at a fixed rate be given for a
much longer period without jeopardizing the effectiveness of the law, at least as to that piece of land?

The land reform is not only not the sole reform, but
it presents questions and difficulties at least as grave
as that of any other well-grounded movement. It
seems from my present light on the subject that the
single-tax, to be practicable, must be levied so as to
leave some exchange value to land; or so that the state
may absorb all interest and rent actually charged on
land and improvements. Unused land may be taxed as
highly as improved land in the same locality, and the
speculation in land and extensive monopoly of this necessity to existence would be greatly ameliorated. It
would furnish an economical and just means of revenue,
settling the tariff question once for all. On these
grounds it is amply deserving of the intelligent, devoted
support which it is receiving. But do not set it up as a
panacea. It is a law-reform, the efficacy of which will
depend on its administration, nothing more.

It is true that advocates whose zeal outruns their judgment announce in sounding sentences that all wealth
might be destroyed from the earth and people would
tap the earth and create it again; implying that they
would do this without serious inconvenience. This statement is so utterly extravagant as to be virtually false.

What was left of the inhabitants of the earth after
such a calamity as the destruction of all wealth would

tap the earth and re-create enough to support them, but what would be left? Bands of savages in the warm zones of continents, where fish and game and wild fruit are abundant. All questions of civilization would go back several centuries and men would work out their salvation anew. This is not the question confronting economists and sociologists to-day. We wrestle with the problem of making life more satisfactory for the teeming, increasing millions now on earth.

If any one doubts the truth of the above view of the reformer's statement, let him divest himself of all wealth and friends, and induce a body of mortals like himself to do likewise, and without even the traditional fig-leaf, let them turn themselves loose on bare land, and refusing assistance from all, "tap it" and revel in the flow of luxuries during a cold northern winter. But this is making too much of a puerile statement. The moral, however, is that in our present state our garnered wealth is scarcely less necessary than land, and the control of that wealth enables one to collect interest just as it does the landlord to collect rent.

If rent were the only source of great and dangerous wealth those who own land of the largest value should be the wealthiest and most dangerously inclined. As a class farmers are comparatively the heaviest land owners; yet, as a class, they are the poorest citizens. They own one-third of the land and yet receive but about one-sixth of the product of the country's industries.

It is entirely certain that anything which would destroy the exchange value of land would make these values attach themselves to other forms of wealth, and without some better industrial system than the present, what would be saved in rent would be paid in interest.

It is as important that the currency and system of exchange should be such that every man should get all the benefit of the wealth he labored for as an individual or produced as part of a great civilization, as it is that the land laws should be such that each should get a share of the land values he helps to produce. If he were allowed to keep what he produces as an individual, the first requisite would be satisfied.

Land reformers maintain, with reason, that men do not produce land and hence it cannot be theirs. That it belongs to men individually or collectively in usufruct only and hence an absolute title thereto cannot be acquired. It is the birthright of all the generations of men. How then can the payment of rent to the state give absolute title to gold or silver, iron, coal, copper, timber or a thousand other things limited in quantity? There is no generic difference in that respect between these and land. The process of reasoning which relegates land monopoly to the dark ages proclaims that at least all forms of wealth, the natural basis for which is incapable of reproduction and limited in quantity, can be claimed in usufruct only and that one man cannot justly charge another for their use.

There are others of an advanced turn of mind who see in postal savings banks a cure for financial ills. How short-sighted! If the postal savings bank does not give the depositor interest it will have an insignificant patronage. Other banks will. If it does, then the body of usurers has been increased, and the mud sills further oppressed. What have postal savings banks to do with the establishment of a rational currency or the prevention of interest-taking?

CHAPTER XXXIV.

The thistle cannot be destroyed by cutting down and scattering its seeds on the ground.

HERE is a means by which interest-taking, the com-mercial incubus of the modern world, may be removed. Destroy interest-taking and all men might work to-gether in harmony, each receiving his own. In a com-munity where no hoarded fortune could last longer than a generation, all would be obliged to work. When each would be obliged to work for what he got, men would soon find that they could work to better advan-tage in unison than each one for himself. Capital would be combined and the best adapted for each task would find his proper sphere. Great companies of toilers working together like bees in a hive would make pro-duction more effective than ever before. There would be no clash of interest as at present, no preying of one on another; he who advanced his own interests must advance the interests of all. The bitter hate and dark contempt of the toiler for his master and the master for the toiler would cease to be. The feverish problems of industrial rights would become clearer.

We would hear no more of the menace and injustice of large fortunes. Cease to allow the Astors to collect rent and interest, and in five years their power would have fallen to the level of ordinary citizens. Their fortunes could not play tyrant for them. Their wealth would disappear slowly but surely until, sometime, that family must take up the burdens and live the lives of other mortals. They would then soon sympathize with others' needs. As soon as interest-taking was made impossible, toilers would cease to pay attention to the holders of large fortunes. There would be neither disposition nor necessity for disturbing old fortunes. Wealth would rapidly accumulate in the hands of toilers, and idlers would be branded with the pauper's stamp. The pauper millionaire would become an extinct species, like the buffalo and the bear. The march of civilization would have stamped him out. No matter how shrewd, or unscrupulous, or avaricious the money-getter might be, he would, without the aid of rent or interest-taking, be utterly powerless to oppress any one by the force of the wealth which he might accumulate.

It is often said nowadays that if the wealth of the world were evenly divided it would within a short time again accumulate in the hands of the same favored few. Granting that there is truth in the statement, it proves nothing except that our laws are unjust. Leave the laws as they are and the unscrupulous, avaricious schemer will usually get the fat of the land. Put in force equitable laws of distribution and the differences in fortune will represent merely difference in ability to produce. Those entitled to wealth would have it. All would work and nobody would be obliged to toil excessively. Relieved of the leisure-bred devices of extravagant ease, our lives would become more rational and simple, and the frivolous exactions which now weigh so heavily on our time and resources would quite disappear. All who were willing to toil would have leisure for recreation and improvement. Art, letters, science, might be cultivated as pastimes by all having such inclinations. We would not have one man with an abnormal development of brain working among ten thousand dunces who could not understand his thoughts. All would be cultivated and intelligent. Remove the incubus of want or fear of want from man and he would be comparatively an archangel. Before the gaze of such a being, crime, like fallen man, would slink away

ashamed.   Three pairs of hands would dispose of what
one wrestles with to-day and the task would be light
to each.   Panics would be a thing of the past.   Depression could follow natural calamity only and the
vigorous country would soon recover itself from the
severest natural strain.   Give the producer the full
measure of what he produces and a giant stride will
have been made toward making the earth what it is
intended to be, a pleasant abiding place for man.   Then
and then only can we have the happiness of the fullest
development and enjoyment of all natural powers.

   This is not a fanciful picture.   It is not claimed that
the reform suggested will make the earth a paradise,
but it will make it more nearly one than it has ever
yet been.   It will lay one true foundation principle on
which future fabrics of civilization may rest secure.   It
will be transferring our social edifice from its foundation of sand to one of solid rock.

   While we give such immense advantage to possessors
of surplus wealth, all men will strive to amass a surplus by any means, however dishonest.   Men have long
since learned that in the present order of things no one
can ever become wealthy by his own efforts in production.   The secret of wealth, the philosopher's stone of
the modern world, is known to be the appropriation by
one man of the results of the toil of hundreds.   It is
nonsense to say that a fortune of a million can be
amassed in any other way.   One might produce in any
tangible line of industry for ten lifetimes and still not
produce to the value of a million.   How to save what
he earns is not now the study of the man of affairs,
but how to obtain legally the earnings of others.   Every
business man's aim under such a system is necessarily
to take every advantage of his neighbor which will give
himself the better of the bargain.   This is "business."
Destroy the law by which man is enabled to appropriate the results of the toil of his fellow-men and you
remove not only his power but his motive for working
injustice. If required to rely for fortune on what he himself produced he would turn his attention to production,
not to filching from his fellows.   A race of human

tigers, jackals and asses would be transformed into a community of soulful, thinking men and women.

There is another side to the picture. We are face to face with the problem of the race; viz., whether the many were born to be the hewers of wood and drawers of water of the few. Civilization has not yet begun to solve it. It has been put off for us by broad continents and virgin resources to subdue and enjoy. We have now reached the length of our tether. We are thrown back upon ourselves. Our country's prestige as a promised land is ebbing. The rights of humanity are at stake. We must decide what they are and how they shall be preserved.

Invention, the handmaid of industry, has been beguiled to the ministry of mammon. We are plunging forward to the plutocratic goal on the wings of steam and electricity. Crises and panics have become as regular as eclipses.

For those who believe in manhood, in the right of all men to life, liberty and the pursuit of happiness, unshackled by the privileges of tyrants and minions of self-assumed privilege; for those who deny that any other man has a right to rule over him without his consent, or that any one has a right to take what he produces without giving him just return—in a word, for all freemen, there is but one course to pursue. Go back to the eternal laws of right and there lay the foundations for your constitutions. Take the truths proclaimed by nature as your polar star and let these guide you to the goal of justice between man and man.

The most important natural truths, with regard to the body industrial, are: That wealth alone supplies the tangible wants of man; that no wealth produces itself; that he who produces it is primarily entitled to the wealth which he himself produces; that, therefore, he is not primarily entitled to that which any one else produces; that the laborer produces all wealth; that the wealth which he produces is, in itself, not only non-productive but comprehends the natural principle of decay; that, therefore, it is advantageous to its possessor for consumption only, and its possession can not be made the basis of a valid claim for any portion of what any one else produces; that the laborer, therefore, is primarily entitled to all wealth. I have applied these principles.

Not one of these truths will harmonize with interest-taking. Interest-taking must be eliminated before an industrial system founded upon true principles can be established.

I have pointed out the way. The plan is feasible, the principles sound. Examine them, and if you find them contrary to truth reject them, but be sure of your case before you decide. The principles are not occult. Go to the task with a love for truth and you can understand. If you find the principles sound make it your aim as a freeman to establish them. It is in your power. You must do it all yourself. The vulture thrives on destruction. A great body of our people, like them, are growing fat on the industrial wrecks about them. They will oppose change. They will be specious and cunning and adroit. They will ridicule and rant. They have you by the throat. The power and organization are theirs. They will press you hard, but if the freeman strikes home, if he goes to the foundation and builds on truth, no corrupt power can overthrow him.

The rank and file of toilers must work out their own salvation. All experience indicates that one having an advantage will never give it up of his own accord. It is those, then, who suffer by the present industrial organization who must bring about a change for the better. Let each fair-minded citizen convince himself and then go ahead and see that his convictions are put into practical form.* It may be necessary to make a little compromise on detail, but on principle, never. In the cause of right, he who wavers is lost, and it is better for the community to make the change thorough-going at once. What would one think of a surgeon who would cut off a limb piecemeal, gangrened though it be, for

---

*To the reformer who is really in earnest there never was a better opportunity than now. Men are breaking away from party control and the set of men who will organize a party on truly democratic principles is sure, sooner or later, to carry the day. If they want initiative and referendum let them first introduce these principles in the government of the political party which advocates such reform. Destroy dictatorship in party, and in the nation your work is half accomplished.

the purpose of sparing the patient suffering? With the stumbling, timid citizen the case is the same. He wastes his energies in trifles which bring no tangible results until the country becomes so tired of tinkering that it casts aside himself and his reforms.

This has been the fatal mistake of tariff reformers. If they had done what the people authorized them to do, this country would be rid to-day of the idiotic spirit of narrow provincialism which has attenuated the great Chinese Empire into a second dotage. International intercourse is the only civilizer yet discovered. Instead of that, the majority of citizens are proclaiming the beauties of exclusion and protection from the house-tops. They are advocating a principle in economics infinitely more absurd than the physical fallacy of perpetual motion.

Let not the same mistake be made with currency reform. Free coinage of silver, real bimetallism, if once completely carried into effect in a rational way, would probably be a considerable improvement on gold, but the relief afforded by such a measure would be entirely disproportionate to the pains required to bring it about. It would be another case of the mountain and the mouse. The citizen would down the gold-bug for the silver beetle, and the "pale drudge" would soon be found to be as inexorable and brutal a master as the yellow tyrant which it had succeeded. A change in the financial system is inevitable. It is the burning question of the day. When it comes let it be such that those who work and suffer for it can point to its fruits with pride and satisfaction. The change must be for the benefit of the whole people. The only change in the financial system which will benefit the whole people is one which will destroy the monopoly in money, take it from the hands of the usurer, and make it the instrument of exchange accessible to all who have wealth to exchange. It must be a currency which will put all men as nearly as may be on an equal footing. It must make it as easy for the producer of any other wealth as for the producer of silver or gold to get a dollar. This may be accomplished by the proposed system. It is the only system by which it may be accomplished. I submit it for the consideration of all thinking, honest citizens.

I do not insist on details, but I maintain that any sound system of currency must be issued on wealth for sale alone, must be issued for all wealth in the process of exchange, must be issued to any one holding exchangable wealth and furnishing security. That the standard must depend on the labor unit and not on the price of any commodity, and that the money token must have substantially no value. I insist further that the holder of currency must be required to bear the burden of care and deterioration of wealth. Above all, I insist that the currency and banking system must be conducted by the government only, in such a manner as to destroy interest-taking.

It is in such a reform and this alone that the toiler must look for a betterment of his condition. There is no other reform which will take its place. It will not take the place of all reforms, but it is vastly more important than any other yet proposed. It lays the ax to the root of the evil.

Now come forward, all ye usurers and money sharks, and sleek, fat, oily bank presidents and officers of loan and trust companies, and tell with divine unctuousness how the proposed currency measures would ruin the country and oppress the toiler. Come forward beaming with an ineffable light of pure disinterestedness and charity unadulterate, and oppose it because it would ruin the laborer and small shopkeeper. Of course you have never given a thought to its effect on your own pocketbooks and privileges. The altruist in you will not admit of that. You who in the extreme goodness of your souls lend for six per cent when you cannot get twelve, and make a thousand by driving some unfortunate to the wall when you cannot get advantage sufficient to make a million, should be listened to as divine oracles of the interests of the rank and file. And the business of shaving notes and exacting usury and engineering corners gives you such a monopoly of political and economic knowledge that he would be rash indeed who would question your inspired conclusions, especially when he knows the source of inspiration.

Oh, no! the people can never understand finances, we could not ask them to try while we have such an extremely well-paid and well-informed body of moneyed aristocrats to give the government all the advice it wants as to the financial needs of the people! And it has been always so clearly shown that a man will work against his own interests to further the interests of some other person. And this is so characteristic of the usurer.

It would be little short of a calamity if the people should come to understand the financial question, for they are always so perverse that their understanding of things differs materially from the understanding of their millionaire friends, and this would be no exception.

And you millionaires who serve your country by drinking foreign wine, attending horse-races and debauching, come up and tell of your divine right to be a charge on the public. Tell of the injustice of making you work like other men. Point out the particular day and date that your father consummated the theft of a railroad and thereby secured luxurious, idle existence for his posterity for all time. Tell how harassed you have been in passing time away, and really how beastly dull it is eking out existence with nothing useful to do. Tell of your trials and worries as though such a tale of woe were a reason why such a state of affairs should continue.

Then tell us how useful your foreign or domestic pleasure-seeking was to the toilers who sweat three hundred days in the year; how necessary it is to have a castle or a hunting park in Scotland in order to carry on business in America; show how, in spending your time in pleasure, you were giving such sage management to business thousands of miles away that your services were really indispensable to the toilers of the land, and without you really they could have accomplished nothing. Tell whether it was yacht-racing or banqueting or gambling which won for you the titles of captains of industry, and define the duties of that particular rank. It would be a long and interesting story, no doubt, for the millionaire to make clear his usefulness to the world which pays him tribute, and give an account of the right he has to what others produce.

Then it would be so interesting for some of our scions of wealth to analyze the motive of philanthropy, which prompts the man who has all that millions can buy to try to make his name and the name of his family immortal by erecting himself a monument in endowing institutions of learning within whose walls plutocracy must find its stronghold with money wrung from overworked, ignorant and degraded laborers. Institutions controlled by millionaires where the fair goddess of knowledge is prostituted to the lusts of mammon, and truth is crucified to protect and perpetuate the ill-gotten wealth of unscrupulous plutocratic tyrants. That is, no doubt, a high type of philanthropy which seeks selfish ends under the cloak of public good. It is morally so much better to give in charity even what we wrongfully get than to leave what we give in the hands of those to whom it belongs, and allow them to use it as they find most advantageous. It is such a balm to the filcher to know that the wealth would be wasted if he had not taken it and cared for it, for himself.

Then the women of ease might take the rostrum and tell, the dear things, how their dreadful sisters would persist in refusing to take from the wealthy shoulders every earthly burden and allow God's chosen creatures to give themselves up body and soul to social vanities; to become the puppets of men of wealth or slaves to a senseless fad. What a frightful world this would be if every one with health and strength were obliged to help himself or herself and if one one-hundredth of the people could not look down in benign complacency on the other mass of dolts and lackeys and give them their pitiful scorn! The rest of the people were really made as a sort of instrument for developing and amusing these chosen ones, and anything which would change such a state of things would make life entirely stale, flat and unprofitable for the chosen few. As for the mud sills themselves, they do not count.

And you learned Solons who have grown gray in pulse-feeling and declamation, ye hoary temporizers who will give the people as little as possible, and proclaim that that little is a boon from the goodness of your souls, let your voices be raised in defense of "honest money," the glittering food of Midas, the shekels of the Shylock. Parrot platitudes which you do not take the trouble to understand. Conjure up the grim shades of Hamilton and Chase, who preached gold and used paper in its worst form. Re-echo the sentences of the long line of bankers and money-lenders who have fully half a dozen times by the power of that gold brought the country to the verge of bankruptcy. Put on the brakes firmly and show that a shining yellow toy is the only instrument which civilized man can devise to carry on his trade with his fellow-men. You will need all of your resources, for the struggle of freedom with their golden fetters is at hand.

Let the Rothschilds and their Cockney confreres come forward and save the dear toilers of America from ruin. Let these usurers who for centuries have drawn a steady stream of wealth into England from all quarters of the world, by the aid of the gold insanity, tell how dangerous it would be to detach the tentacles of the gold octopus from the industries of the world. Let them point out the magnitude of the calamity of ceasing to pay billions each year to the syndicate in the control of the gold of the world, for the control of that gold. Yours will be an interesting story. But every citizen of the country has a stake as dear as you—his life and happiness. Do not assume that you alone have a right to be heard.*

All you who thrive by taking part of what he produces from your laboring brother, will be expected to fight a rational currency system to the death. You have cunning and knowledge and the power it gives, but the waves of popular discontent are surging against the fetters of iniquity and wrong. When the pressure

---

*I do not hold for a moment that the wealthy are not personally as virtuous as the poor, and, under like circumstances, would be governed by like motives; but it would be flying in the face of all experience to expect them to voluntarily give up the industrial advantages which the present order gives them.

becomes too great the golden bands will go with the rest, and we will awaken after the deluge to a regenerated and revivified industrial system. Interest-taking is being tried as never before, and this time the inquisitor is intelligence instead of superstition. It cannot fail to fall, and on its ruins must be built the new and just financial system.

CHAPTER XXXV.

"Saw a vision of the world and all the wonder that would be."

I SEE before me a country fair to look upon. Its grassy slopes are animate with herds and flocks, its smiling valleys wave with wheat and corn. Its broad rivers stretch away to the great ocean, bearing upon their bosoms wealth untold. The sun-kissed spires of great cities stud the banks and spread away over the gilded hills. Mighty lakes float a commerce never dreamed of by Carthage, Sidon or Tyre. The oils and silks and spices of the East mingle in lavish abundance with the meat and corn of the West, in bursting palaces of wealth. Rugged mountains lift their giant forms and shake from their dripping flanks the refreshing dews of heaven upon the teeming life of the mellow plains and hills; or oppose their broad backs to the storm king as he rages in destructive fury upon their precious charges which sleep beneath the sun. The ample strong-boxes of those rugged mountain kings are filled with treasure of iron and lead and coal and silver and gold and oil, and the hale, generous guardians toss back the keys and invite the millions of the earth to help themselves.

Glittering bands of steel knit together ocean and lake and river, mountain, hill, valley, dale and woodland. The locomotive, the shuttle of industry, flies to and fro weaving the warp and woof of national prosperity. It is a scene fit to gladden the hearts of archangels.

Again I look upon the enchanting scene and I see it

250

instinct with life. About the teeming valleys and sunny
hills are seen the images of God above. I draw closer.
There is an intense earnestness in those pinched and
sun-browned faces, shaded by sere and withered locks.
Grim, wiry forms in somber-hued rags bespattered with
dust and grime, move to and fro. Thin, bony fingers
clasp, with the iron grasp of toiling brawn, curiously
wrought contrivances of many shapes. The dull, som-
ber eyes gleam with resolution as the lines move along.
It is a battalion of the army of toil assaulting the
treasure-house of mother Nature   The scorching sun
directs his burning shafts full upon the hosts, but they
flinch not. Great beads stand on their throbbing fore-
heads as they stare the flaming monarch full in the eye,
but they waver not, nor pause in the earnest strife. The
heavy ax resounds, the ploughshare grinds the pebbly
soil, toppling forests crash, the garnering sheaves send
out their furnace glow, but not a line is broken. The
grim column marches on.

A cloud of anxiety darkens each sunburnt brow.
Some far-off goal never to be attained, seems dimly in
sight. Each casts a furtive glance behind, then
startled, turns and toils toward some distant goal. I
look and wonder what is the prospect which gives the
toiler's eyes that dull, eager look, and at his shoulder
I perceive a dim, formless thing, the ghastly specter of
want. As the toiler pauses in the arduous conflict, I
see this hideous monster raise his skeleton arm and
point menacingly his bony finger to a great chill mansion beyond the hill, then with a leer, to the sought-for
distant goal.

The dull eyes of the toiler forever wander there.

All about the chill mansion across the hill are groups
of men and women whom the weight of toilful years is
bearing to the earth. The knotted limbs but half bear
up the shrunken bodies, and the locks, whitened by the
frosts of time, toss about over faces from which the
light of hope has fled, and left a blank waste or pathetic second childhood. It is there, to the almshouse,
the specter points and leeringly contrasts it with the
cottage of peace and plenty beyond, always far beyond.

The eager eyes strain harder, the toiling muscles knot,
and the struggle with nature for competence is pursued
with redoubled force. There is no rest, no stopping for
breath.   Ever present by the toiler's side is the grim
specter to remind him of the almshouse or the cot of
peace. He groans as he sees his fellow soldiers drop
their worn weapons from their enfeebled hands and
shamble off in the tottering procession to the chill
mansion across the hill.

My eye wanders from the open, sun-kissed fields and
the grim army struggling there, to the spire-decked,
mansion-studded cities. There is a great black building
where the turbid river tide chafes against the murky
bank. Its walls are grimed and smoke-stained and
its windows thick and gray. I look within. A band
of giants are gathered there, stripped as for a mighty
conflict. The tense muscles knot on the sturdy bare
arms and chests. Great drops of sweat stand upon their
brows. Their faces are dull and sullen as they gleam
with a weird distinctness in the fierce white glare of
the great hot furnace. Its dull radiance lights up the
great rafters and soot-festooned nooks, and lends a sort
of unearthly glow to the whole dark scene. This is
another battalion of toil, meeting face to face and sub-
duing for the cause of man the tyrannous hosts of nature.

The same eager look suffuses each pale, smoke-grimed
face. The same grim resolution, softened now and
then by indifference or reckless insouciance. The same
goal is far ahead, the same grim specter at each shoulder, the same bony finger and the same cunning leer.
But close beside the gloomy mansion across the stream,
surrounded by the wrecks of toil and years, are the cold
gray walls of a great chill building through whose narrow barred windows the sunlight never falls on the de-
fiant, scowling faces within. And as the giant of the
furnace follows with his glance the specter's bony finger
along the dusty road beyond, he sees an unkempt, be-
sotted face staring from shaggy locks, and a form
scarcely human staggering along in a mass of filthy
rags,—the tramp, the derelict of the highways of earth.
The specter seems to laugh a low laugh of satisfaction

at the scene, but the furnace's fiery eye has fewer terrors for him than the scene without, and the giant plunges again into the thought-devouring conflict.

Dazed, I turn from the scene and look around for a fairer prospect. But wherever I go I see beneath the surface the same grim, unrelenting conflict. In the sightless depths of the dark mine, in the balsam-laden forest glades, on the rugged mountain slopes, in the broad harbors, upon the tall ships, behind the flying locomotive, in the stony, sun-parched streets, in the sickly, nauseating tenement, in the stifling, overcrowded factory, I see battalions of the same army pursuing the same relentless strife. The same sordid eagerness lights up the eyes; the same fixed determination, the same despair, hold sway by turns over the stolid features. The same, whether the face be the pinched and sun-browned features of the son or daughter of the soil, the bleached cheek of the sturdy giant of the furnace or shop, or the delicate brow of the pale girl of the factory or tenement.

Why this terrible, unending, hopeless strife, embittered by the unbroken presence of the grim specter of want? Can such a promising exterior of beauty, peace and plenty, hide but care, woe, dread and disappointment? Why this unrelenting, hopeless toil, why this pathetic ending? Why the almshouse, the prison, or the inclement road, with its menacing human derelicts, as a reward for years of toil? Where then is beauty and happiness? Is this fair earth but a chamber of horrors masked with a glittering tapestry?

My eyes again wander along animated streets of fair cities, where tall piles of steel and brass and stone and marble rise. Within close but sunlit rooms, I see pale faces pouring over broad books. That same look of eagerness and care is to be seen, but a cunning light lends vivacity to the face. The restless eyes glance furtively over the pages as the brows knit and the form sways to and fro. The specter is entirely absent or but dimly visible, but the eager eyes still strain toward some coveted object beyond, a formless thing encased in golden robes bedecked with jewels. It is a deity

never met face to face, but known to fame and fable as
the Goddess of Success.   I am interested.   I follow the
pale-visaged votary as he descends in the iron car and
rolls noisily away to the broad avenue, beyond the
work-a-day turmoil of the city streets.   I follow him
up the marble steps into the tapestried and cushioned
drawing-room.   I follow him even to the banquet
board, I hear the glasses clink and see the red wine
flow.   I hear the rustle of silks and see the forms of
delicate women flit about under the mellow glow of the
electric lamps.   I hear merry voices and mirthful
laughter, but in it all is a minor note of discord, a
harsh, grating, sordid sound.   That eager gaze, still,
distorted features, even flushed with wine.   The laughter pealed and rolled and rung, but more and more
became a senseless, hollow, mocking sound, an almost
frightful thing.   The looked-for happiness was not there.
And I found it not on the wave-kissed sands, among the
free, unconventional crowds of merry-makers; not in
the salon, nor the yacht, nor the theater, nor the lec-
ture hall, nor at the concert, nor the church, nor the
ball; but instead the same hollow laugh of unreal mirth,
the false, jarring note, the same eager look.   The specter of sham was as constantly present among the battalions of fashion and pleasure as was the specter of
want among the battalions of toil.   The chamber of
horrors was still there, even though the gilding was a
little more ingenious.   I looked and listened and asked
why.

An imperative desire took possession of me to know
whence came all this cruel fear and bitter unreality in
a land of beauty and plenty.   I watched and saw again
the army of toil plunged into the contest, the enemy
the formless thing of yore.   Fast and sullenly did the
toiling army pile between them and the bony monster,
Want, stores of the thing men call wealth.   It seemed
the only barrier between them and the dominion of the
monster's rule.   As they turned to wrest from nature
still more and more, to raise the barrier higher, other
brawny giants step boldly up, and seizing from the
barrier whatever they desire, bear this wealth away to-

ward the palaces of pleasure and the mansions of the
pale-faced men with the cunning eyes. · These pale-faced
men watched and directed that it be put away to ap-
pease the all-devouring goddess of luxury and success.
But even the brawny men who bore this store of wealth
to the shrine of the gilded goddess kept but a little to
protect themselves, and the specter was also ever at their
backs. I saw now and then a grim-faced toiler look up
and see his painfully accumulated store disappear behind the strong doors of his pale-faced neighbors. He
would look at the messenger of the votaries of success
coming back always empty-handed for more, and scowl
until the bony finger of the specter would remind the
man who stopped to think, of the grim horrors of his
domain, and the toiler would sullenly bow his head
and toil again. I saw a cordon of sturdy men, the
strongest of the land, drawn about the store-houses of
the pale-faced, cunning men, and the rations of these
guards were also taken from the meager barrier which
the soldier of the battalion of toil placed between himself and want.

There seemed an unwonted movement in the toiling
ranks. A few bolder than the rest dared to stare the
specter in the face and watch as their stores were transferred to the houses of the votaries of pleasure and success. Soon they noticed that not one of those pale-faced,
cunning men took part in the battle against the common enemy, Want, but bent all of their cunning energies
to carrying away and appropriating the barriers which
the battalions of toil reared against the implacable foe.
Even those who did the work of taking away the wealth
were toilers like themselves.

First they stood dazed by the revelation, awed by
their disappearing defenses, then with a courage born
of desperation they called to their companions to look.
The knotted sinews relaxed and the dull eyes opened
wide. The pale-faced, cunning men pointed to the
specter who hissed in low, chuckling tones: "Toil or
perish." The whole chamber of horrors was displayed
before the toiler's eyes. The almshouse and the prison

and the derelict-haunted road, even to the nameless
grave in the potter's field. As the battalion leaned upon
its arms the bony finger lifted the veil. There before
him appeared the suffocating tenement where chill
death painted black shadows around eyes gleaming
with fever. Parched infant tongues, furred and hot,
panted in vain for the saving draught. Haggard, faint-
ing mothers tottered empty-breasted over dying babes.
Cold, dank basements, reeking with filth, sheltered men
turned to brutes and children transformed to savages.
The whole panorama of distress is seen; the pale-faced,
cunning man leers at each detail, but still the grim
hosts of toil stop and look. I hear a low voice hiss,
"Toil or perish!" but the host moves not. There is a
desperate gleam in eager eyes and a dark scowl on each
sullen face as they see their hard wrought store disappear and leave them to the mercies of the monster Want.

"By what right?" arises in a hoarse whisper and is
caught up until it becomes a mighty chorus which sends
a thrill of fear through those of the cunning eyes and
makes their pale faces darken. "By the right and authority of the law of which these are the conservators,"
hiss the pale-faces with a sneer, and they point to the
cordon of blue-coated giants who stand between the
toilers and their fast-disappearing stores. Grim mutiny
appears in every haggard toiler's face. If the law is
against them, their case is desperate indeed. Soon I
hear a murmur rise, "Let us change the law, the law is
ours." It is caught up slowly at first, then faster and
faster by the sullen ranks, until it becomes the general
war-cry of the whole host.

The cunning eye becomes sinister, the soft-handed
captains argue, then threaten, then plead. They call
upon the blue-coated giants to preserve inviolate the
rights of property. They point out to these that without such hoards to protect, a large part of their occupation would be gone. But the blue-coats remain stolid.
The power was then on the other side and it would be
folly to try to stem such a current. They could not if
they would. There was still one resort: the gladiator
battalions whose business it is to kill.

But they too were foiled. No law had been violated, it had been but changed, and the chances of a conflict against such a force with law on its side were not the best. The soldier did not budge.

There are menacing faces in the ranks of toil, hot indignation goads these hardy men to violence as they see the pale-faced, cunning men plotting to defeat the law, but calm-browed brothers step before their wrath and admonish them to patience. "The law is ours. We have no need of what is stored away in their possession. Let us keep our weapons, the instruments of toil which we now use, and let them have the rest. Let the dead past bury its dead."

The pale faced, cunning ones stand aghast. They see the whole world of industry move along without the slightest relation to them. They see men of their class fast taking places in the ranks, glad to have their wealth used and preserved for them, and their toil remunerated, simply as the toil of others. The stores of toil rise high, and the grim specter slinks away. His course is even to the domain of pleasure and success, even to the abiding-places of the bediamonded goddess on her gilded throne, whom men were so wont to worship. Even the old menace of the specter of want is turned toward the frivolous and gay of yesterday, as the stores of toil disappear from their palaces. The chamber of horrors now exists for them. The pale-faced, cunning man is no longer serene in his security. His appeals to police and soldier are alike in vain. He sees the blue cordon which formerly existed for him, wheel about and bar him from the stores which the toilers gathered. Then one by one those sturdy conservators of the law disappeared into the toiling ranks.

Then the pale-faced, cunning man grew gracious. Yes, he would give to the legions of toil for their use all that he had if they would but give him the increase. He was met by a merry, hearty laugh. "Increase indeed! that is a fiction of other days. 'Tis toil, and toil alone which increases; come, be one of us," and instruments of toil are placed in the soft, white hands. Slowly, reluctantly, he takes up the fight against the now disappearing specter of want, the specter who now hovered near the palaces on the hill.

I looked and even the terrors of the horror chamber were disappearing. The pinched, sullen faces of the days gone by had brightened. They were softer and had lost some of that eager look. The incubus of the dread of want had lifted. There was no specter constantly menacing those who worked. No visions of the almshouse or the potter's field, of famishing babes and starving mothers. The mine, the shop, the forest and the field were filled with rosy, beaming faces The faces in the sunlit offices are not so pale, the eyes not so hard and cunning The throbbing trains bear their precious burdens of human freight, but it is not the sordid worshipers of sham. The gleaming shores of the broad oceans have their crowds, but what a change! The shrines of luxury, frivolity, pride, licentiousness had been transformed into the Meccas of mirth-loving, healthful toilers on happy holiday outings. The palaces of sham and mammon had become palaces of art. The earth had its bright fruits and flowers and balmy winds and grateful suns alike for all. The fair exterior of placid streams and sunny hills and teeming valleys and rugged mountain slopes were but the outward expressions of the peace and happiness within. It was the happy home of a transformed man.

THE END.

www.ingramcontent.com/pod-product-compliance
Lightning Source LLC
Chambersburg PA
CBHW021522210326

41599CB00012B/1349